A brilliant and funny tour through mythology, evolution and the day-to-day innovations of scientific research, this is an entomological page-turner. 'If there were no dung beetles,' Byrne and Lunn write, 'there might have been no human race ... They literally change the earth beneath us.' This book reveals that earthly transformation in fascinating and lucid detail.
— **Bruce Beasley, Professor of English,**
Western Washington University

Biology and history dance with the scarabs in this beautiful book with its wide-ranging perspectives on our changing understanding and appreciation of these marvellous creatures.
— **Jane Carruthers, Emeritus Professor, Department of History,**
University of South Africa

... some of the most insightful research on dung beetle behaviour is dealt with in this beautifully written and illustrated book. It is a fitting tribute to these remarkable insects and to the authors who have written about them in a scientifically profound yet charmingly simple way.
— **Clarke Scholtz, Emeritus Professor of Entomology,**
Department of Zoology & Entomology, University of Pretoria

Dance of the Dung Beetles *shows the delightful and charming side of the dung beetle enthusiast ... scientifically rigorous and highly readable!*
— **Sandra Swart, Professor of History, University of Stellenbosch**

Marcus Byrne is Professor in the School of Animal, Plant and Environmental Science at the University of the Witwatersrand, Johannesburg and has studied dung beetles for more than 30 years.

Helen Lunn has a PhD in Musicology and has a wide research base. She has worked in academic and popular writing environments.

Dance of the
Dung Beetles

THEIR ROLE IN OUR CHANGING WORLD

MARCUS BYRNE AND HELEN LUNN

WITS UNIVERSITY PRESS

Published in South Africa by:

Wits University Press

1 Jan Smuts Avenue

Johannesburg 2001

www.witspress.co.za

First published 2019

http://dx.doi.org.10.18772/12019042347

978-1-77614-234-7 (Paperback)

978-1-77614-235-4 (Web PDF)

978-1-77614-236-1 (EPUB)

978-1-77614-274-3 (Mobi)

Copyeditor: Helen Moffett

Proofreader: Steve Anderson

Indexer: Marlene Burger

Cover design: Hybrid Creative

Page design: Hybrid Creative

Typesetter: Newgen

Typeset in 11 point Crimson

To our families

Acknowledgements

A WORK THAT EXPLORES DUNG beetles across a seven-thousand-year time span is by definition ambitious. The level of that ambition only became apparent towards the end of the task – probably a good thing, as we might otherwise never have attempted to write this book.

After having dipped into fields ranging from Egyptology to evolutionary biology, a sense of our own shortcomings as writers led us to rely on the expertise and kindness of many specialists. The generosity of the responses of those we approached for comment was heart-warming and enormously helpful. The academic world is filled with experts in specialist fields, and stepping into such terrain can be a nerve-wracking and dangerous affair. Because we skim over areas to which scholars have devoted years of research, there are many sins of omission and commission, along with our obstinate interpretation of the literature. We take full responsibility for these sins, but we would like to thank all of the following for their assistance and very useful observations and commentary.

We are grateful to Salima Ikram, for advice on ancient Egypt, and Claudia Tocco for checking our facts on Ulisse Aldrovandi. Norman Owen-Smith is thanked for his insights on elephant research in Tsavo National Park. Eeva Furmann, widow of Ilkka Hanski, provided details of her late husband's research at Oxford University. Clarke Scholtz looked over our South African dung beetle accounts, while Penny Edwards gently corrected our errors and lapses of memory surrounding the events and outcomes of the Australian Dung Beetle Project. Marie Dacke and James Foster straightened out our misdirected ideas on beetle orientation and vision, while Leigh Simmons applied valuable selective pressure to the chapter on evolution. Our thanks go as well to John Meyer, who translated a section of Aldrovandi's *De animalibus insectis* for us. Chris Collingridge shared his illuminating pictures of dung beetle research. Beautiful beetle pictures were loaned by Adrian Bailey and Shaun Forgie. Thanks are also due to Wits University Press and our publisher Roshan Cader for patience and gentle guidance through the process of producing a book, and delightful gratitude goes to Helen Moffett, our editor, for unstinting energy and good humour, loaded with Beatles puns, at all hours of the day and night.

A final thank-you to all the members of the worldwide community of beetle people for your generosity of thought and time.

Introduction

IN THE SUMMER OF 2009, one of us (Byrne), at least, was having fun. He was in the bush with his friends, playing with dung beetles. These friends, a group of scientists from Sweden, Australia, Germany and South Africa have managed to get together every year since 2003 to run experiments on dung beetle orientation.

We had already shown that dung beetles were the first animals known to be able to orientate by polarised light from the moon. Our next task was to measure how the nocturnal species performed when compared with their diurnal (day-active) counterparts. This involved working all day and most of the night when the moon was in a particular phase, getting slimmer as it waned into a silver sliver lying on its back in the African sky. We were tired but happy. The nocturnal beetles were incredible; they could roll their dung balls in a straight line under a cat's whisker of a moon.

But when the moon was absent and we relaxed, drinking cold beer under the light of the Milky Way, we were fixated with the sky. If we could see this ethereal light, then surely the beetles could too, and therefore use it for orientation? At the time only humans, along with a few species of birds and seals, were known to be able to orientate by the stars. Our beetle companions were a (relatively) large, enigmatic ball-rolling species called *Scarabaeus satyrus*. We knew they could do it, but needed to prove it with scientific rigour. The key challenge was to stop the beetles looking at the sky, which is equivalent to asking a goldfish not to swim. How does one stop a beetle looking at the sky? Not so difficult if you fit it with a little peaked cap. But not that easy either when it has no ears to hook things onto, and its head is flat and shiny and has evolved to stop anything sticking to it.

Nevertheless, once their hats were glued precariously in place, the capped beetles were lost, wandering aimlessly with their dung balls, all dressed up and nowhere to go. Ten minutes earlier, the same uncapped individual had streaked across the starlit savanna with the confidence of a taxi driver heading for home. Problem solved – the beetles obviously needed to see the sky to find their way around – but which bit? The night sky, even without the moon, is a complex conglomerate of constantly moving light. Unpicking which part of the sky the beetles needed for pathfinding would reveal the compass cues they were using. This suggested that we should take them on a trip to the planetarium.

Night after night, the dependable beetles enthusiastically pushed dung balls around the Johannesburg planetarium, responding to the pinpricks of light reflected from the domed ceiling with gusto. By systematically removing elements of our ersatz night sky, we were able to conclude that the Milky Way, the very centre of our own galaxy, is part of their nocturnal compass. Consider that a dung beetle hatches inside a cocoon of dung deep underground and spends most of its life grubbing around eating, moulding and transforming faeces beneath the earth; it is nothing short of inspiring to learn that the brilliant band of light that is the Milky Way is the beetle's reference point when it plots a path through the African night.

The dung beetle has a miniscule brain, much of it devoted to analysing smells, and yet it can process visual information that even humans with their vast brains struggle to comprehend. The contrast between the little insect and the immensity of its visual references reminds us in an oblique way of Hans Christian Andersen's story about a dung beetle.

It's a charming story about a rather arrogant little dung beetle who lives in an emperor's stables. The emperor's horse is honoured with golden shoes as a result of saving the emperor's life in battle. Upon seeing his neighbour the horse shod with golden shoes, the dung beetle sticks out his skinny little legs and demands golden shoes. The farrier declares him out of his senses, to which the dung beetle replies that he is equally a resident of the royal stables and therefore as worthy as the emperor's horse.

The farrier absolutely refuses to provide the golden shoes and so the beetle quits the stables in high dudgeon. This triggers a series of adventures in which the dung beetle fails to find either his place in the world or his own satisfactory identity. He does not really learn humility but when he finally flies home to the emperor's stables, landing exhausted on the back of the favoured horse, he reasons that the golden shoes on the horse were actually put there to honour him, the dung beetle. And so he is reconciled to his home. When we consider how vast our little planetarium dung beetle's view of the galaxy is, it might feel that it well deserves those golden shoes after all.

As a group of scientists working on the spectacular adaptations of everything from the neural pathways in dung beetles' brains to their orientation and navigation abilities, we were contributing to an ever-increasing body of knowledge on the complexity and variety found in this large family of very successful insects. When we published our Milky Way results, the story was picked up way beyond the scientific literature and spread rapidly around the world. We were struck by how the story of a lowly dung beetle and the immense Milky Way engaged popular imagination when so much other information about dung beetles is equally impressive, if not even more fascinating. This realisation prompted us to respond on behalf of these little creatures (which can be found on every continent except Antarctica) to show that they deserve better press than to be seen as mere dung-grubbers, some of whom happen to orientate by the stars.

Dung beetles appear in early creation myths, and go on to take a central role amongst the Egyptians and their beliefs about life and death. Dung beetles' fortunes have fallen and risen with the shift from a world dominated by a religion that symbolically incorporated them in some of its key concepts of rebirth, to a world in which science has largely separated itself from religion. The ascendancy of the dung beetle in our collective knowledge mirrors the development of science during the past five centuries: how it moved from being a quest by individuals trying to substantiate the biblical version of creation, to a pastime of well-heeled gentlemen in solitary pursuit of knowledge, to the dazzling but modest brilliance of Charles Darwin, and on to a collective of scientists engaging with increasing volumes of detailed data on animal and entomological development and behaviour. It is the very nature of these scientific practices that have changed the way we understand and view dung beetles in our world, and their value to us as an important source of information about the way the world works. It is that story, combined with their role in our most cherished beliefs, which turns these remarkable little dung consumers into creatures of immense interest.

This book is the result first of musings, then of discussions between the two authors. We both realised that we could look at some of the history of the development of science in a microscopic way through a focus on dung beetles. They have been ever-present in the history of the West (but oddly, less so elsewhere), in religion, art, literature, science and the environment, and what we understand about them now mirrors our greater understanding of the important role they play in keeping our planet healthy.

This is the story we explore in the following chapters. It is a story with a few unexpected twists, as it moves from the tombs of the pharaohs to the drawing rooms of the directors of the Dutch East India Company to the remote forests of Madagascar. It is a big story carried on the back of a family of small creatures who seldom diverge from their dogged pursuit of dung in its infinite variety and abundant supply. Like the housemaids of Victorian Britain, who tended fires and households in the small hours while the Empire swept across the globe, they remain largely unseen and ignored. Yet without them, the world would have a lot more dirt in it. Dung beetles are largely invisible, and yet without their vital activities the world would have a lot more faeces in it.

One thing we find, looking back at the history of our understanding of dung beetles, is a progression in the way we as humans have viewed the world. We started with magical thinking about the world around us. From there, we moved into symbolic ways of explaining the world, and then rational thought (which has ebbed and flowed) came to dominate. That progression was neither linear nor sequential. We continue to retain magical thinking in the way we address the world around us, though overlaid by rational thought; symbols blending with reason. However, by exploring this one creature and its fortunes we can see how our ideas and practices surrounding our relationship to the world have changed. This is what we set out to do in the following chapters as we follow the fortunes of dung beetles in human history.

CHAPTER ONE

When the dung beetle wore golden shoes

DEATH IS NOT A SUBJECT one expects to find at the beginning of a book on dung beetles. The idea of someone's great-great granny wandering around with dead dung beetles dangling from her ears is equally strange, but the two subjects are not unrelated. The Victorians in their grand obsession with Egypt, death and loss shared a number of ideas with the ancient Egyptians from whom they took the association of dung beetles with death. The difference in the case of the Victorians was that they had a monotheistic religious template for death, which differed from the Egyptians' rich animist pantheon of gods. Moreover, instead of wearing scarabs made out of stone, the Victorians frequently wore the real thing. Quite how the hapless beetles found themselves adorning the earlobes of respectable ladies is part of a story that began seven thousand years ago in Egypt, and which came full circle with the nineteenth-century invasion of Egypt by Napoléon and the subsequent development of Egyptomania. It was the two subjects of death and resurrection that made dung beetles so significant in ancient Egypt.

Although the family of dung beetles comes in a huge array of sizes, bizarre shapes and iridescent colours (with some so small you can barely see them) it is the smaller subfamily of true dung beetles that earned these insects their central role in Egypt. What made them so important to the Egyptians was their intimate relationship with dung, which promoted them to godliness. This makes the idea of dainty/fastidious Victorian women boldly wearing such creatures as ornaments seem even odder. The Victorians, however, viewed dung beetles and nature in general as a window onto the mystery of creation and as a distraction from the ugliness of industrial society, so perhaps they did not consider the faecal associations of their entomological jewellery too closely. The same cannot be said for the ancient Egyptians, who were very aware of the dung-rolling proclivities of the beetles. It is this improbable and intriguing relationship between an insect known for its relationship to 'filth' and the beliefs of one of the most enduring civilisations known to humankind that is our point of departure.

Why would a creature that subsists on the most unappealing end product of other living creatures be the one insect (as opposed to so many others) singled out for sacred recognition? And why primarily the ball-rolling species? On the other side of the world, the early culture of the Paraguayan Lengua-Maskoy people also gave ball-rolling dung beetles a mythological role. The dung balls seem to have been the clincher almost every time; this is not completely surprising, given that the balls of fresh faeces the beetles sculpt in full view are usually so beautifully spherical and (relative to the size of the beetle), evidence of remarkable strength. Although the handsome rollers are the most obvious component of the dung beetle fauna, only about ten per cent of tropical dung beetles actually roll balls, and that percentage declines as one moves towards the poles. The other species are either dung dwellers (like the beautiful *Oniticellus formosus*) or tunnellers (which are by far the largest contingent of the dung beetle fauna). Tunnellers commute between the dung/soil interface and their underground nest, so they are rarely seen unless one is prepared to poke through poo. It seems, however, that only a few species of rollers in the huge subfamily of over six thousand species lie at the core of the myths and beliefs that surround these beetles. It is their visible burying of the balls of dung in the earth (together with a seemingly miraculous reappearance from that same earth) that earned the beetles singular significance in certain early mythologies and belief systems.

Dung beetles and their balls have made sufficient impression on humans to feature in creation myths on more than one continent. Among the Bushongo people of the Democratic Republic of the Congo in Africa, the scarab was seen as the original insect which created all other insects. This is an unusual variation on other creation myths, in which the scarab's dung-rolling activities usually feature more prominently. In South America the Chaca Indians (who, along with the Lengua-Maskoy, lived in the area of present-day Columbia) gave their dung rollers a singularly important role. For them the dung beetle was the potter, a giant beetle known as Aksak, who created nothing less than man and woman.[1] This creation myth is still present among Indian tribes resident in Bolivia's Gran Chaco country, but has no resurrection counterpoint as found in Egyptian mythology. Instead, relatives generally destroy memories of their dead, who are believed to turn into animals – except for the shamans whose souls live among the stars of the Milky Way. For the Toba of Sumatra, the dung ball symbolised the ball of matter the scarab brought from the sky in order to form the world. Clearly it was the moulding of the dung ball that was symbolically important to the Chaca and the Toba. This gives us some understanding as to why that little ball has managed to carry so much significance.

Dung is inescapably dung, however, and in other cultures it was not so much the ball as the beetle's relationship to dung that was of importance. This earned dung beetles a place in a Chinese Taoist text, *The Secret of the Golden Flower*, which was based on ancient oral transmissions of an esoteric Chinese circle.[2] In *The Secret of the Golden Flower* the dung beetle was a metaphor for the transformation central to the Taoist concern with being at one with nature. This offered the means of transcending the self in order to be at one with the world. The dung ball, from which life would eventually develop, was an example of how the spirit might grow and transcend the environment in which an individual lived. This theme recurred in the writings of Christian theologians, albeit in the context of dung as a metaphor for sin. Unlike the Christian tradition, the Tao grew out of a form of animism, the beginning point for most spiritual expressions for early humankind.

While dung beetles seem unlikely role players in our complex faiths and the history of human belief systems, they were far from alone in feeding the human imagination about answers to the origins of life and the hereafter. Natural resources (in particular certain mushrooms and other psychotropic plants) were responsible for visions of other worlds beyond the immediately visible one, and fed conceptions of realms that were somehow linked to the material world. Dung beetles entered these alternative worlds simply by rolling and burying balls of dung and thus unintentionally mimicking the daily death and rebirth of the sun, the first deity.

The combination of invisible realms with the symbolism of an insect that seemed to replicate the motion of the earth is as intriguing as the raft of beliefs which followed. Both gave early humans visions and symbols, along with ways of translating the seeming chaos and uncertainty of life on the planet into digestible human interpretations and ways of thinking. The fungi and mind-altering plants consumed by the early shamanic mediums shifted their consciousness into shapes, forms and connections that defied words, but which were compelling enough that they were (and still are) believed to represent an unseen and significant reality. Meanwhile dung beetles, through their dung burial behaviour, became symbols of beliefs surrounding death and resurrection: the passage of day into night and back into day, from light into dark, and back into the light again.

It was probably shamanic journeys into both the under- and overworld that fed the earliest myths of what we now call heaven and hell. The Mayan text, the *Chilam Balam of Chumayel*, is one of the few manuscripts to have survived the destructive judgement of Catholic monks in South America – they burned anything that they believed referenced pagan beliefs. This almost impenetrable text, which appears to be a mix of Christian cosmogony (theories about the beginnings of the universe) and traditional Mayan beliefs, might have escaped destruction for this very reason. In this work the scarab appears as the filth of the earth, in both a material and moral sense. As such, it represented something similar to early ideas of hell as a place and feeling of primal chaos, disorder, sickness and pain, in contrast to heaven as the realm of expansion, healing and release.[3] In ancient times heaven was a world accessed either through rhythmic drumming and trance dancing, through psychoactive plants or through a combination of the two. Along with dreams, those experiences were the earliest happenings that created a sense of unseen worlds, equally numinous in their inclusion of the mystery of where people disappeared to after death.

Although the dung beetle does not appear to have any place in modern mythological belief systems, the early pre-Egyptians saw them as a powerful symbol of resurrection and the creative power of the sun. It is via this route that their legacy was embedded in every major Western religion, from the time that humans first developed spiritual curiosity about where we came from, and to where we are going.

The gradual desertification of the Sahara (which began to accelerate approximately 5300 BCE) motivated small groups of nomadic hunter-gatherers to cluster increasingly around the Nile River, where they began to farm its banks and to found one of the earliest and most enduring civilisations in human history. These nascent societies created an initial foundation for the multiplicity of gods that we now identify with early Egyptian civilisation.

Death and order were central topics in these developing civilisations. In their late Neolithic burial sites at places such as Deir Tasa, there is evidence of a belief in an afterlife, with the dead being buried with bowls of food and other items thought essential to life in the hereafter.

They also established an assemblage of gods and beings they had relied on in smaller tribal groupings. As animists they appear to have seen the cosmos as a 'Grand Chinese Opera', populated with a limitless cast of frogs and elephants, rivers and mountains and probably beetles.[4] Over time the pantheon of Egyptian deities incorporated everything from crocodiles and bulls to dung beetles.

Whatever early belief system one examines, it is not surprising that animals present in the local habitat were the ones invested with meaning and significant roles. The Japanese used the cicada as their symbol of resurrection in the hereafter. Life-giving animals such as the auroch (a forerunner of the cow which offered food, clothing and hides for warmth) were particularly significant in the cold climes of Europe. For the San people of Southern Africa the eland was an equivalent creature, a particularly powerful beast whose death opened up a space in which the living (through the intercession of a shaman) could connect with other realms and the spirit world.

In Egypt, the mystical animal-deities were those still seen in tomb paintings and sculptures. They ranged from fierce or dangerous creatures such as lions, crocodiles and cobras to harmless birds and insects. Different groups revered different creatures, and the ruler of each district in pre-dynastic Egypt had his own god, which explains how a number of important deities (such as Horus, the falcon god) were introduced. The beginning of dynastic Egypt circa 3100 BCE saw the consolidation of regions under one overarching ruler, a Pharaoh (king) who, like the rulers in Egypt's pre-history, identified with one or more animal gods.

One of the defining differences between a Pharaoh and a shaman was that the latter was usually an otherwise unremarkable community member within a small social and kinship grouping, usually consisting of no more than 150 people. The shamans became priests and part of a formalised religious hierarchy, a radical contrast to the shamanic world of the egalitarian hunter-gatherer, in whose animist universe all creatures and things were able to address each other. The priest now became the intermediary to the gods, who were relied on to ensure rain and bountiful harvests. This meant the evolution of a powerful priesthood, in which the Pharaoh was the chief mediator with the gods, with the priests below them in the hierarchy. The role of the priests was to care for and mediate with the gods, who in turn would show their pleasure through the annual inundation of the Nile and plentiful harvests.

Over time priests, priestesses and god wives (wives or consorts of the Pharaoh) became enormously wealthy and powerful. As they ostensibly communicated with the gods rather than the people, a language representing the gods in the form of symbols helped to cement the concept of different deities. Shared symbols with the same meaning for all were required for group cohesion, and represented a step away from the imagery and objects used in the earlier smaller groups. Dung beetles would have been well known around the rim of the Mediterranean, with the sacred scarab, *Scarabaeus sacer*, occurring in Southern Europe and North Africa; the term 'scarab' is used as a common name for dung beetles to this day. The sacred scarab eventually rose to such popularity in ancient Egypt that, 2 500 years after its zenith, both Pliny and Plutarch commented on the beetles' surprisingly high status.

The formalised appearance of symbolic scarabs as amulets in the early Egyptian Middle Kingdom from at least 3000 BCE suggests scarabs were already a well-established representation drawn from the natural world at the beginning of dynastic Egypt.[5] Dung beetles were embodied in the god Khepri, a creature presented with a human body and a scarab head. His name is variously transliterated in English as Khepra, Khepera or Khepri meaning 'itself is transforming.' He was a magical creature, a perception reflected in the play on the name for dung beetle in Egyptian, 'hprr', which means 'rising from, coming into being' and which became 'hpri', the divine name Khepri: 'It is said of Khepra, as of Horus, that [he] produced the Ma, that is the law or harmony which upholds the universe.'[6] Khepri is a significant god because he was associated with creation and becoming, whether in this world or the next. He was a special deity from the beginning of dynastic Egypt (circa 3100 BCE) and much of his fame was put down to his particular (but misinterpreted) breeding behaviour, in which it was assumed no females were involved.

It has also been suggested that the jewel-like colours of many scarabs might have first attracted human attention (as many beautiful species still do) resulting in dresses for movie stars and grandiose ballroom ceilings being decorated by their metallic wing covers. (Prada handbags, designed by the mercurial Damien Hirst, continue this eccentric and impractical but nevertheless beautiful fashion on the runways of Europe). However, the sacred scarab (which is generally agreed upon as the model for Khepri) is not quite as colourful as some of its other scarab relatives, so brilliance is not a particularly compelling argument in this case. Its significance, since its first appearance as far back as the second prehistoric civilisation in the Nile region, derived from its ball rolling and breeding habits, and the interpretations of these behaviours.

Dung beetles were placed in jars buried alongside the dead, indicating that their ability to disappear into the earth and then re-emerge was a magical attribute that could influence the fate of the corpse. Both dried beetles and models of beetles made out of green serpentine and cut sard were found in graves at Tarkhan. Scarabs and other beetles were considered sacred and magical from the earlier part of the second prehistoric age in Egypt right down to the Christian period.[7] Khepri's role in Egyptian belief systems increased in significance as the pantheon of gods and their roles evolved. A correlation between the number of gods and the increasing complexity of the newly settled society is apparent, but Khepri was included from the start; his fortunes and representation (although not static) remain in one form or another throughout the very long history of Egyptian culture.

In Memphis, the ancient capital of Egypt, Khepri was associated with Ptah, who was originally the god of the earth from which it was believed humans were crafted. It was this association of creation and shaping forms from the raw material of the earth that linked him to Khepri, whose dung ball represented the daily passage of the sun and the coming into being of matter. Khepri also featured in one of the many Egyptian creation myths in which a lotus flower rises out of the waters. When its petals open, a scarab is revealed which transforms itself into a weeping boy; the boy's tears become humankind.[8] This potency was also associated with gods who represented the subjugation of chaotic natural forces, particularly with the dwarf figure of Pataikos, on whose head he is depicted. Khepri's name was often included as one of the five great names of the King that reinforced the message of the King's relationship to the gods.

Science in early Egypt, a necessary outgrowth of agriculture, was an attempt to explain the way the world worked, and involved mathematics and geometry. These intellectual tools were used to lay out and record ownership of land, which frequently disappeared under the life-giving silt deposits that arrived with the annual flooding of the Nile River. The sciences were used to establish a calendar based on that critical event, which was vital to the local agricultural practices and ultimately to the survival of society. From this we can assume with some confidence that the Egyptians observed dung beetles pretty closely, something confirmed by the fact that they made a distinction between the scarab that went underground to breed (which was associated with Osiris, god of the dead) and the young scarab which emerged from the ground (associated with the son of Osiris, Horus the falcon god). They presumably noted the ravages of age on the parental beetle, with its cuticle dulled and front legs worn down from digging over three years of adult life. In contrast, the newly emerged adult would reflect a metallic sheen to match its pert spikey new legs as it broke free from its earthen chamber into the sun.

Unfortunately, we have only this indirect evidence of early Egyptian naturalists excavating the ground where they might have seen a dung beetle disappear, but even if they followed him underground, they still subscribed to a magical and symbolic explanation for the significance of his re-emergence. To the Egyptians, Re (Ra) was the chief god while Horus was 'the one far above'; so when a combination of Horus's wings and tail with the newly emerged young Khepri began to appear inside the royal pyramids at the end of the fifth dynasty, this new representation cemented Khepri in his role as the symbol of resurrection. This was a key concept for the Egyptians, who believed that life did not end with death. Even though we do not know the exact origin of this belief, which was certainly not limited to the Egyptians, this does appear to be the first time we have actual documentation of it appearing in the records of a civilisation. It is a belief still widely held in the contemporary world, and represents a core concept in the development of religious perceptions around death.

Although we find mention of Khepri in the Pyramid Texts, which are dated around 2350 BCE, he was at his peak of popularity during the period known as the New Kingdom (1550 BCE–712 BCE). This was characterised by five hundred years of exceptional prosperity, during which the pharaohs amassed huge wealth from the extension of Egyptian influence into the near Middle East and south into parts of Nubia (present-day Sudan). Much of this wealth was used to build elaborate temples and to construct statues dedicated to Egyptian gods. Nowhere was this more in evidence than at the temple complex at Karnak, which was constructed from around 2055 BCE, but which in the period of the New Kingdom became one of the most important temples in all of Egypt. Eighteenth dynasty rulers, in particular Thutmose III (1479 BCE–1425 BCE), began a series of ambitious embellishments to the temple complex, resulting in it becoming one of the largest religious sites ever constructed. Thutmose III's reign is considered to be the greatest age of the scarab, when its use as a symbol was most common and most varied.

In the Karnak temple compound a huge red granite statue of Khepri, dedicated by Amenhotep III (1386 BCE–1349 BCE), was moved at some point to the northern corner of a sacred lake. This position was imbued with significance, as the human-made lake symbolised the primeval waters from which it was believed all life arose; Khepri crawling out of the life-giving mud after the Nile receded from the flood plains was associated with the coming into being and rebirth of life and the sun. The positioning of the Khepri statue at the northern side of the lake served as a symbolic representation of this relationship; ironically, the symbolism of this positioning was not significant enough to prevent the Egyptian Antiquities Authority from moving the statue from its initial position to ensure a better flow of tourists at the site. The statue rests on top of a cylindrical pedestal, the front of which has been flattened to form a *stela* (a carved or inscribed stone slab or pillar used for ceremonial purposes) on which is carved a kneeling king offering two vessels to Atum (the primordial god who arose from the sacred waters). The text on the pedestal reads 'Khepri who rises from the earth'.[9]

Few things could express Khepri's importance to ancient Egypt more visibly than this huge beetle statue. However, his journey from those lofty heights to the present is one that traces our changing understanding and perception of the world: his name now survives mainly as a classical reference in modern entomology, with *Kheper* referring to a group of over twenty (mainly Afro-tropical) dung beetle species of great beauty; such species in the sub-genus *Kheper* are the archetypal ball rollers and members of a larger grouping, the genus *Scarabaeus*, which includes the sacred scarab *Scarabaeus sacer*.[10]

Different species of scarab amulets were crafted during different dynasties, and the inscriptions on the undersides of these scarabs have helped us to reconstruct some of the history of ancient Egypt.

In dynastic Egypt, for example, Khepri not only symbolised the belief that death was not the end of life, and that resurrection was real; he was also a metaphor for the passage of time and the movement of the sun through the course of the day. Everything about the life of the ball-rolling dung beetle was interpreted as portentous. It rolled the ball backwards – a peculiar way for any animal to move – creating a living image of reversible life. The scarab itself at times becomes the sun, because it flies on hot clear days. Most importantly, the beetle created a circular ball that disappeared underground and from which a new dung beetle emerged. For the Egyptians, this internment of the dung ball was akin to the disappearance of the sun at night and its reappearance every morning. The dung ball thus became a symbol of the sun. It was believed that the setting of the sun represented the separation of the body and the soul of the sun, and that when the sun rose the following day, the two were reunited. To the Egyptians, who observed nature closely, this seeming capacity for self-regeneration was the transformative ability that promoted the dung beetle into a truly auspicious symbol.

As a deity, however, Khepri represented more than a belief in resurrection. He was also identified as a primal source of creation. It was this that distinguished Khepri from other creatures also thought to be representative of the proof of resurrection. For the ancient Egyptians, anything that disappeared and then reappeared was included in the pantheon of creatures representing the concept of rebirth. Because they submerged themselves in water to resurface some time later, hippos were also associated with regeneration and rebirth, but the dung beetle and his ball spoke of life eternal while the hippo had a more ambivalent set of associations. Hippos were dangerous and associated with chaos, while female hippos were associated with mothers and childbirth in the shape of the goddess Taweret. The fearsomely mythical devourer of souls who failed the feather test at their judgement after death was Ammit, an amalgam of a crocodile, a lion and a hippo, all animals known to kill humans. The activities of dung beetles, however, always remained benign or beneficial.

We can deduce from all of the above that dung beetles were closely studied, but we also know that their early observers suffered some confusion over what they were seeing. They believed that the male dung beetle was capable of parthenogenesis (that it had no need of a female to reproduce) and that this was an additional quality that made the dung beetle supernatural. The Egyptian assumptions about Khepri can to some extent be explained by the fact that the sexes of ball-rolling dung beetles are not that easy to distinguish, and that females of many insect species (including dung beetles) can store sperm between matings, precluding the need for the constant attentions of a mate. In addition, the dung ball rolled across the ground does not contain an egg: this is only added days later, after the female has reworked the ball underground, often on her own. If we consider that in Western science, the individual roles played by egg and sperm in the creation of a new individual were not finally settled until 1876, such shortcomings in Egyptian dung beetle biology shouldn't be treated with too much scorn. They do, however, give us a clear marker of the limits of observation by the unaided eye – something which would define natural history for centuries. Based on their assumptions, the Egyptians concluded that there were no females in the entire scarab family. Inscribed on the stela of Hor and Suty, two overseers of works erected during the eighteenth dynasty, we find confirmation of this belief in the lines:

Great Hawk, dapple plumed;
Scarab who raised himself up;
Who creates himself unbegotten.[11]

Less easily confirmed is Plutarch's assertion that Egyptian warriors had a beetle carved on their signets as a symbol of masculine bravery as well as immortality, an assertion which appears to be myth transformed into fact. A more nuanced understanding of the role of males and females in ancient Egypt suggests that in some way Khepri, when seen in relation to Maat (the concept of balance), embodied the duality of male and female. Considering that he was a magical creature with mythical attributes, his ability to procreate in the absence of females was not all that implausible.

Different periods and towns in Egypt had differing beliefs about the formation of the world, but Khepri was an enduringly popular god who managed to survive many changes in the expanding Egyptian world: bringing good fortune and (most importantly) ensuring the life-giving and sustaining sun rose every day, just as a well-preserved body would eventually rise in the future. Various Pharaohs owned beautiful and valuable armbands and amulets representing Khepri. Less elevated individuals possessed simpler representations of him, but his presence was pervasive in both life and death, granting protection from evil in life and the promise of resurrection after death. The Egyptians had scarab amulets inscribed with everything from good wishes to deities' names, and well-worn personal scarabs were usually found with the mummified remains of their owner.

The Pharaoh Amenhotep III, who ruled from 1386 BCE–1349 BCE used such scarabs to reinforce proclamations about his own personal greatness and the extent of his power. Surviving scarabs from his period boast of his having killed 102 lions. Other scarabs promoted the name of his wife Tiy and his dominion over Lower and Upper Egypt, as well as his prowess in capturing wild cattle. Amenhotep apparently understood the value of self-promotion in public relations and the power of the symbolic, in which Khepri was a notable vehicle for transmitting his message.

Yves Cambefort, a renowned French entomologist and dung beetle specialist, has conducted extensive research into the role of scarab beetles in early cultures. He suggests reasons for the elaborate and complex rituals surrounding Egyptian mummification can be found in the idea of regeneration via the dung beetles' brood ball, its signature means of reproduction. Cambefort has proposed that the Pharaoh's sarcophagi reference the pupal chamber into which a dung beetle larva eventually transforms its brood ball, a suitable environment for the possibility of rebirth of the body.[12] According to his theory, the immobile pupa represented the mummified body. The Egyptians believed that a person's body was necessary for resurrection, hence the need for it to be elaborately preserved (including the bandaging, which mimicked the segmentation of the inert beetle pupa). This

14

interpretation is based on the assumption that at some point, someone dug up a brood ball and investigated its contents, which is entirely possible. While we have no evidence other than conjecture for this, it is nevertheless a tempting conclusion.

Egyptian mummification progressed over time, signifying the importance attached to the physical body of a person. Depending on the wealth of the individual who had died, mummification developed into a refined art ensuring that the deceased person would take their body, considered as a requisite for the next life, with them. As the art of mummification became more sophisticated, all the major organs were preserved and dedicated to the four sons of Horus: the liver was protected by Imsety, Hapy was associated with the lungs, Duamutef with the stomach, and Qebehsenuef with the intestines. The only organ not removed from the body was the heart, which was considered to be the seat of intelligence. Small scarabs were usually placed upon the eyes or the breast and sometimes over the stomach, a practice which varied according to the time period. The most important scarab, however, was the heart scarab of the mummy, which was much larger than the other scarabs and frequently made out of basalt or green stone. This was a form of insurance to protect the deceased at the decisive time when their heart would be weighed against a feather by the funerary jackal god Anubis before a panel of 42 judging deities at the threshold of the afterlife. Judgement of the goodness of their heart was critical to their chance of resurrection. The underside of the amulet was usually inscribed with a prayer, beseeching the heart not to bear witness against its owner in the final judgement.

Cambefort has extended his theory about mummification and dung beetles further by suggesting that the interiors of some of the remarkable pyramids built to house the Pharaohs bear a curious similarity to the chambers and tunnels created by dung beetles underground; this again suggests that the Egyptians might well have done some digging to find out where dung beetles went underground. The structure of some pyramid interiors resembles the steep narrow passageway down to the beetle's brood chamber, which houses the brood ball in which the dung beetle larva grows. Intermediate compartments (used for temporary dung storage), turns and blind-ended passages between the surface and the final chamber add to this similarity. Cambefort suggests that encasing the body in a 'brood ball' inside an elaborate and impressive structure that closely mirrored the underground home of Khepri ensured resurrection of the pharaoh: doubly important because pharaohs were considered to be gods who, once they died, would be reunited with Osiris, the god of the afterlife, life, death, resurrection and transformation.

These theories are not improbable, but they do raise some questions; if these funerary practices and monuments were based on the model of the dung beetles' underground life, why did the Egyptians build some pyramids above ground rather than underground? And why do we find pyramids in other cultures where dung beetles were not gods? Cambefort's ideas are intriguing, if speculative, but one answer to the above-ground building of pyramids lies in the belief of the ancient Egyptians that the continuation of the life of a person after their death was connected to people remembering him or her. A solitary soul was undoubtedly dead, while a person with many people in their life was alive. One of the virtues of a vast pyramid was that it kept the dead pharaoh alive in people's memory; and as long as they remembered him, he lived on.

Hieroglyphics (*mdw netjer* in ancient Egyptian, translated as 'the words of God') were an important part of the representation of the Egyptian world view. The text of Horapollo Niliacus, written in the fourth century CE, provided an early explanation of ancient Egyptian hieroglyphs. Although the authenticity or accuracy of much of what Horapollo wrote is disputed, the work is generally accepted as being written by someone who lived when hieroglyphic signs were still in use. Horapollo wrote that the Egyptians believed that the dung ball was also a symbol of the world, which in turn suggests that they were aware that the earth was round and not flat. They observed and recorded a great deal about their world: they knew that a month was on average 30 days long, and they overlaid this knowledge with that of the brood cycle of the dung ball, since it took 28 days for a new beetle to emerge from a ball of dung (though in reality this would vary with the temperature and rainfall of the region and the species of beetle involved).[13] They believed that this timing corresponded with the passage of the earth through each of the twelve signs of the zodiac. When the dung beetle emerged from the brood ball on the 29th day, they believed that this coincided with the initiation of a new regenerative cycle of the earth.

Examining the world of the Egyptians reveals a search for order and attempts to make sense of their world. They looked closely at the night sky and the passage of the planets, and saw patterns, along with answers found in those patterns. They developed sophisticated ways of observing the stars and planets, and many of their observations persist in modern astrology. The heavenly Egyptian scarab remains in the modern zodiac as Cancer, metamorphosed by the Greeks and Romans into the crab. The symbolic presence of the persistent scarab in the stars adds romance to the finding that at least one species of nocturnal ball-rolling dung beetle can use the Milky Way to orientate its ball-rolling. To paraphrase Sir David Attenborough, had the ancient Egyptians known that the lowly scarab beetle could use the centre of our galaxy for orientation, they would surely have felt vindicated in elevating it to the status of a god?

Horapollo tells us that Egyptian observers noted how the dung beetle pushed its ball with its hind legs, head down, going backwards 'rolling its ball to the West, while himself looking towards the East'[14]. This is possibly one of the earliest written records of animal behaviour, linking the beetle's rolling direction to the position of the sun. It wasn't until the work of Ulisse Aldrovandi, a remarkable Italian naturalist intimately acquainted with Egypt's most recondite mythology, was published in 1602 that the mystery of dung beetle reproduction was finally unravelled. It took almost another four hundred years to explain the influence of the sun on the orientation behaviour of dung beetles, and we now know that the sun is but one celestial cue (but nevertheless the primary one) used by dung beetles to plot a path for their dung ball through unknown terrain to a burial spot. The Egyptian commentary would have required a close study of their own dung beetle fauna, but we have no record of who made the original observations of the beetles, nor how they came to their conclusions – which were largely correct concerning the influence of the sun in the orientation behaviour of dung beetles. However, Egyptian interest in dung beetles was never really a pragmatic one. Although the dung beetle was connected to the life-giving period after inundation by the Nile (when the beetles would reappear as harbingers of growth), over a period of more than three thousand years the dung beetle left the realm of the mundane and became firmly entrenched as a symbol of Egyptian optimism in the capacity of humans to transform and regenerate themselves.

The Egyptians' belief in life after death was shared by the early Christians in the same way that Egyptian practices and beliefs have left their traces in many other contemporary religions. The practise of circumcision was an Egyptian custom which appears to have been adopted by Judaism, as was the eschewing of pork in any form. However the centrality of resurrection in the Christian religion is of particular significance, as it lies at the heart of the ethical and moral foundation of that church. The notion of Jesus Christ dying for humanity's sins and being resurrected from his grave after three days in the tomb is absolutely central to the Christian narrative.

We have no way of proving exactly how core spiritual concepts moved between groups of early modern people, and there is always the danger of claiming that if one society had an idea, other cultures that share a similar idea must have copied it. Similar ideas can arise repeatedly and independently, with monotheisms appearing in human cultures which we know to have been separated by major oceans for most of their history. It remains noteworthy, nevertheless, that the key notions of burial and rebirth are so similar in Christianity and Egyptian mythology. The fact that Khepri as a symbol did travel outside of the borders of Egypt supports the theory that some aspects of his symbolism might well have influenced other cultures.

We find Khepri in the discovery of scarab amulets in the Middle East and in countries around the Mediterranean basin, most of which traded with Egypt: Sardinia and Crete have ample evidence of this traffic. The Phoenicians borrowed the scarab from Egypt, and after they were overrun by Shalmaneser III, King of Assyria (860–825 BCE), scarabs found their way to Nineveh and Babylon. Scarabs have also been found in the graves of the Philistines in present-day Israel.[15]

We cannot assume that the spread of the scarab as symbol would have conveyed the exact beliefs it represented for Egyptians. Yet it is unlikely that it would have travelled as far and wide as it did if some vestige of its reputed symbolic power had not accompanied it. At an archaeological site in Mesopotamia in present-day Syria, scarabs were found at the temple of Ninkarrak, the Mesopotamian goddess of well-being and healing. The scarabs found there were not made in Egypt, but were of Levantine origin; and similar scarabs of non-Egyptian origin have been found at the sites of Sidon, Byblos, Ugant in the northern Levant, and Tell al Ayjul in the Southern Levant. These scarabs were used as amulets and were symbols of good fortune, happiness and beauty.[16] Their presence in graves in particular suggests that some of their original meaning and value might well have been retained and, like other myths, found their way into new and different mythologies about the origin of human beings and the world.

Christianity's formal establishment of monotheism as central to its doctrine meant that all the other gods, particularly those drawn from the realm of insects and other animals, had to be dismissed. Once Christianity took a proper hold around the Mediterranean in the latter days of the Roman Empire, Khepri the dung beetle god had to be erased and forgotten along with all other gods. The influence of Egyptian thought, however, remained. In the pre-Christian era, the Greeks had already added another level to the spiritual journey after death. In their writings, the soul not only could live on after death, but could return to earth reincarnated in another form, a subtle variation on the concept of resurrection. The Greek historian Herodotus (circa 484 BCE–425/413 BCE) states that several Greek authorities adopted a doctrine of transmigration of the soul based on the Egyptian beliefs, and used it as their own.

The Greeks had different symbolic uses for the dung beetle, and in *Aesop's Fables* (which date from approximately 500 BCE), a dung beetle appears as an earthy and dirty creature that can nevertheless fly to the chief god Zeus. The story in *Aesop's Fables* indicates that the dung beetle was independently noticed by the Greeks, and was not influenced by the concept of Khepri. In the fable, an eagle hounds a hare, who begs the dung beetle to protect him. The dung beetle implores the eagle to leave the hare alone. He warns the bird not to ignore his pleas, his small size notwithstanding, and in the name of Zeus to abandon his pursuit of the hare. The eagle takes no notice and proceeds to tear the hare apart. The dung beetle is so incensed that he rushes off and destroys the eagle's eggs. Not content with one year's revenge, he proceeds to do the same the following year. The eagle, desperate to protect his future brood, takes the next lot of eggs to Zeus and places them in the god's lap, trusting that they will be protected there. The indomitable little beetle simply fills itself with dung and flies straight into the face of Zeus, who, startled by the filthy creature, leaps up, dropping and breaking the eagle's eggs. The beetle then relates to Zeus the wrong the eagle did, both to him and to the name of Zeus. The god tries to persuade the dung beetle to relent, but the persistent little insect is not prepared to do so. Zeus, who does not want the race of eagles to expire, arranges instead that the eagle will lay its eggs at a time of the year when the dung beetle is not active. This story is difficult to verify in modern scientific terms, as the species concerned are unknown. However, most dung beetles have a well-defined period of above-ground activity, linked to rainfall and rising temperatures, which would be around the end of winter in the Mediterranean. This may well not overlap with the breeding times of local raptors in the region.

Another Greek writer, Aristophanes (circa 446 BCE–386 BCE), would later incorporate these details into his play *Peace* (written in 421 BCE). This play imaginatively combines a mythical past, history and bawdy humour. The story concerns a shortage of food in Athens and how one father, Trygaeus, proposes to solve the problem by flying to the gods on a dung beetle to ask for bread. His young daughter is incredulous, but he points out that everyone knows from Aesop's tale that dung beetles alone can fly to the abode of the gods. To which she replies:

Little Daughter
Father, father, that's a tale nobody can believe! That such a smelly creature can have gone to the gods.

Trygaeus
It went to have vengeance on the eagle and break its eggs.

Little Daughter
Why not saddle Pegasus? You would have a more tragic appearance in the eyes of the gods.

Trygaeus
Eh! Don't you see, little fool, that then twice the food would be wanted? Whereas my beetle devours again as filth what I have eaten myself.

Little Daughter
And if it fell into the watery depths of the sea, could it escape with its wings?

Trygaeus
Exposing himself.
I am fitted with a rudder in case of need and my Naxos beetle will serve me as a boat.

Little Daughter
And what harbour will you put in at?

Trygaeus
Why, is there not the harbour of Cantharus at the Piraeus?

Little Daughter
Take care not to knock against anything and so fall off into space;
once a cripple, you would be a fit subject for Euripides, who would
put you into a tragedy.

Trygaeus
As the Machine hoists him higher.
I'll see to it. Good-bye! *To the Athenians.* You, for love of whom
I brave these dangers, do ye neither fart nor crap for the space of
three days, for, if, while cleaving the air, my steed should scent
anything, he would fling me head foremost from the summit of
my hopes.[17]

It is highly unlikely that the fables of Aesop or the comedies of
Aristophanes were a consideration among the early Christian teachers
who, emanating from the school of Alexandria in Egypt, were
assembling the critical texts of Christianity. Khepri disappears as a
symbol, but the dung beetle persists, albeit in less elevated roles. Its
descent in some cultures is rapid. In Neo-Platonism, the dung beetle is
stripped of all magic and reduced to the lowest of the low. Porphyry,
a Syrian philosopher (234 CE–305 CE) demonstrated this attitude in
his argument that everything in nature had a purpose, other than the
following hideous and repellent creatures: flies, mosquitoes, bats,
dung beetles, scorpions and vipers. It was inevitable that as the world
became metaphorically flat, the dung beetle (the symbol of resurrection
and a sun god) would crash rather ignominiously.

The dung beetle was now ignored in a world in which, as the Bible proclaimed, humans had dominion over the earth and all its creatures, which in turn became subject to and lesser than humankind. This seems a minor point, as does the dropping of the dung beetle from its symbolic role in ancient Egypt, but it is not. The centuries that followed, in which humans separated their existence into a hierarchy of creation of which they were the apex, were critical to forming subsequent views of the world. The questioning behaviour and philosophies of the Greek thinkers and the school of Alexandria, which had flourished more or less simultaneously, had become dangerous because of their potential for heresy, and were suppressed. The nature and purpose of 'Man' was defined and narrow, and no questioning of why or how we came to be on earth (along with everything else in the natural world) was encouraged. According to the church patriarchs, the Bible provided all the answers and no further enquiry was necessary. Yet, as we shall see in the next chapter, the dung beetle appears in the writing of some of those church fathers in the context of its relationship to dung. Its days as a powerful and independent god, however, were completely forgotten.

Dung beetles must have thrived on the Christian rejection of the cult of the body, as practiced by the Greeks and Romans with their baths and obsession with the human form. Poverty and a lack of interest in the beauty and cleanliness of the body showed one's commitment to the Christian god, and encouraged turning away from concerns of the physical world to a focus on the spiritual. One inevitable outcome was that dirt and disease grew and prospered, and the not very clean world that ensued was a good one for dung beetles, and one in which to some degree they helped to keep humanity relatively healthy through disposal of faeces. We can find evidence of an awareness of this vital role of dung beetles in language, such as the old Anglo-Saxon word for dung beetle: 'tordwifel', or turd weevil, a creature at a long remove from its role as a deity.

With the growth of the priesthood and the Christian church in the West, centred in Rome, knowledge that threatened these power structures was often classed as dangerous and heresy. The majority of congregants lived in ignorance and were unable to read or write; this suited Rome, preoccupied as it was with power struggles and battles with kings and the aristocracy as it expanded its empire in the form of 'holy wars' and Crusades (in which the promise of god-sanctioned plunder proved most attractive). The truly great monuments to the long centuries of the Middle Ages are the awe-inspiring cathedrals of Europe, which reinforce the focus of the period on a trust in and reverence for the Christian god. The soaring spires of those cathedrals advertised their connection to God in his heaven, but ironically it was the celestial explorations of astronomers such as Galileo and Copernicus that gradually nudged the earth from the centre of the universe, to a quiet suburb of the Milky Way – our ordinary galaxy, one of billions.

It seems fitting that nocturnal dung beetles have been looking at the Milky Way for millions of years, understanding its significance as a guide to path finding. It was dissension within the ranks of the Christian church about the accuracy of the scriptures that, when compared to new discoveries in astronomy and anatomy, marked the beginning of the Enlightenment. The advent of the Gutenberg press and the translation of the Bible into the vernacular liberated the Bible from the control of the Church. At the core of the Protestant Reformation was a desire for ordinary people to communicate with God through Jesus, not via a priest's paraphrased Latin scripture. King Henry VIII, in his quarrels and ultimate breach with Rome, ordered a copy of the Coverdale Bible (a precursor of the King James translation into English) to be chained to the pulpit in every church in England. If the parishioners were unable to read it, they now had an incentive to learn.[18] This growing access to increasing information and knowledge began to change the world, and ultimately ended the subjugation of humanity to the belief that the Bible was the only source of all the knowledge required to live on earth.

CHAPTER TWO

Crawling out of the darkness

THE FORTUNES OF DUNG BEETLES in the Middle Ages are relatively muted, and their main use appears to have been in remedies used by peasants. It was at the intersection of the Middle Ages and the Renaissance that the beetles began to reappear in the work of naturalists interrogating the heady mix of religion, alchemical pursuits and traditional knowledge of the previous thousand years. This small band of individuals frequently oscillated between alchemy and observation-based knowledge – hardly exceptional, given that there never is a clear break between historical periods. The word alchemy itself was derived from the Arab word 'alchimia' and referred to magical and mystical traditions combined with investigation into the nature of the physical world; it encompassed the very fluid boundaries between those activities.

Despite official injunctions against magic, magical formulas were as prevalent in medicine as they were in alchemical formulas for transmuting base metals into gold.[1] Such formulas appear in the *materia medica* (collections of information about the therapeutic properties of anything used for healing) of the Middle Ages, scattered among the more prosaic use of flora and, to a lesser extent, fauna. The actual recording of the use of animals and insects in the West can be traced back to Xenocrates of Aphrodisias, a first-century Greek whose work is only indirectly known via the writings of Galen (circa 130–210). Xenocrates's influence spread to the Arab world, and it is in the *Syriac Book of Medicines* that we find a very medieval fate for black scarabs – they were boiled in olive oil to cure earache. Dung itself (without any attendant beetles) featured far more widely in medical concoctions in the Middle Ages, often in the form of bizarre mixtures which worked best (if at all) as placebos.

In China, dung beetles appear in early *materia medica*, and are still used today by practitioners of traditional Chinese medicine. These traditions began in rural areas, where peasants used whatever materials were at hand to effect cures. Some of these proved efficacious, and were recorded. Contemporary research in entomotherapy and compounds shows that different Chinese scarabs and their larvae have potentially useful antioxidant, antifungal, anti-viral and even anti-cancer properties, which probably explains why their use has continued for so long.

The Middle Ages marked a time when scientific development in the East outstripped that of the West; however, apart from a remedial role, dung beetles do not appear to have contributed to the impressive achievements of Eastern scholars. Then came the period known as the Renaissance in the West (from about 1350 to 1600), which coincided roughly with the slowing down of the Chinese and Arabic paths of discovery and experimentation. It was during this exciting historical period that the humble dung beetle became detached from its role in religious mythology and folk medicine. It continued to appear in unexpected places though, not the least of which were the musings of deeply religious but conflicted thinkers who saw the beetles as metaphors for both sin and the common human being. The beetles also crept around the globe (literally) in association with the new voyages of exploration as they unwittingly hitched rides in ships' ballasts.

The dung beetles travelling the world as unwitting colonists were to help reshape the physical world as much as the early explorers circumnavigating the globe. Driven largely by visions of glory, greed and the soon-to-be abandoned belief that their journeys might lead to the location of the original Garden of Eden, the early explorers carried their Christian concepts of the world with them. When Columbus landed in the Americas in 1492, he believed that he had found the Garden of Eden, but the discovery that most of the New World's plants and creatures were not recorded as having been on board the Ark created great confusion. The number of new foods that were different from those in the Old World was extensive, and raised the question of the possibility of a creation different from the one mentioned in the Bible. Moreover, there was neither wine nor wheat for the celebration of Mass in the New World. The Portuguese were to continue the search for the Garden of Eden in South America, but the ultimate realisation that there was no physical Garden of Eden had a profound effect. The imagined world that had sustained both belief and social structures for centuries was starting to crack, and the light of evidence-based observation began to filter through those cracks.

The revelation that revered thinkers such as Aristotle (circa 384 BCE–322 BCE) had been incorrect in their beliefs (making the claim, for example, that the equatorial region was too hot for life) opened up the possibility that if authorities such as Aristotle were wrong on certain subjects, they might have erred on many other fundamental ideas and beliefs. Aristotle's description of dung beetles, when contrasted with some of his other biological observations, illustrates the contradictions presented by his work. He had described the dung beetle as a creature that 'rolls dung into a ball, lies hid in it during the winter, produces small larvae in it, and out of these come more dung beetles.'[2] On the one hand, this vaguely correct observation provides an indication that even in such an early period, evidence-based naturalism had a home in the classical world; but it highlights the inaccuracy of some of his other statements, which included the nonsensical claim that insects were devoid of internal organs. He likewise claimed that men had more teeth than women, and that flies had four legs. The fly error might have come from a misinterpretation of Aristotle's description of a particular mayfly that uses its front legs like antennae; and miscounting teeth might have occurred in a society where people lost teeth easily. Once received wisdom could be disproved by evidence, however, the way opened up for new challenges and methods of examination and discovery.

Nevertheless, these new ways were always going to have to contend with the Bible and its authority, and traversing the sixteenth and seventeenth centuries provides abundant evidence of this tension. The question of humanity's relationship to God and the world was a central question in early Christian writings. It is in there that we find the earliest tradition of dung beetles as metaphors for sin, human weakness and – most surprising of all – Jesus.

The Physiologus, a didactic Christian text written in Alexandria by an unknown author in the second century, described dung beetles as symbols of heresy and their dung balls as evil thoughts. They were rather more kindly depicted in the metaphors of St Augustine (354–430), the Tunisian theologian. Augustine, who was (and remains) one of the most influential early Christian authors, referred to Jesus as his *Scarabaeus sacer*, his 'own good beetle', because he had not only taken on human form, but had been born out of the same filth as humans. This likening of human weakness to the dung of dung beetles proved to be very tenacious.

St Augustine's 'good scarab' is intriguing because as a North African, it was very likely that he had come across Khepri in his home country of Tunis (now part of Algeria) or possibly Carthage, where he studied. He would also have encountered large ball-rolling beetles in the genus *Scarabaeus* or subgenus *Kheper*. Augustine's knowledge of Khepri seems even more likely because he repeats the concept of dung-beetle parthenogenesis, a core feature of Khepri. Unlike the Egyptians, however, he did not see the dung ball as a symbol of the sun, but rather as a symbol of moral turpitude. It is easy to assume that St Augustine derived this metaphor from his knowledge of Khepri, but this is not necessarily the case. St Augustine found his true calling in the Catholic Church only when he travelled to Italy. It was there that he encountered the famous cleric St Ambrose of Milan (340–397), who was to influence him profoundly.

St Ambrose also used the concept of 'the good *Scarabaeus*' numerous times, describing God as 'The good *Scarabaeus* who rolled up before him the hitherto unshapen mud of our bodies.'[3] The mystery surrounding this concept lies in the fact that St Ambrose never travelled to Africa, and that it was Augustine who was influenced by him, not the other way round; how then did the concept of the good *Scarabaeus* enter into a Milanese bishop's thinking? These early Catholic theologians were preoccupied with establishing the Church, and would no doubt have been horrified had they known that their *Scarabaeus* would one day be used by a fifteenth-century theologian to criticise the very structure their writings had helped to cement.

Erasmus of Rotterdam (1469–1536), justifiably known as one of the great humanists of the Northern Renaissance, used scarabs in his *Adagia* (published in 1500) to criticise the abuse of power by the Catholic clergy. In this annotated collection of more than 4 000 Greek and Latin proverbs, the dung beetle first appeared as the champion of the weak, while the eagle hunting him represented the tyrant – the clergy. It was a dangerous reference to make at a time when state and religion were in conflict, and in later editions of Erasmus's *Adagia* the tale became more general and the dung beetle disappeared. Yet Erasmus liked the metaphorical value of dung beetles, and continued to use them in his writing.

Fifteen years after the publication of the *Adagia*, Erasmus transformed the dung beetle from a symbol of the common people to that of a creature little different and yet cleaner than a sinner. In his *De Immensa Dei Misericordia*, he wrote 'What is more vile than the dung beetle? Yet the dung beetle is clean compared to a sinner in his squalor.'[4] Erasmus was not done with the useful dung beetle, whose little hole or burrow he used to frame a question about God's omnipotence. None other than the Protestant reformer Martin Luther (1483–1546) responded to Erasmus's question about whether God was present in the dung beetle's hole, arguing that God was indeed everywhere because he had created everything.

The religious disputes of the time were subtle, but ultimately were always aimed at understanding our role on this earth, and the value of life. The implicit assumption was that we were here thanks to a beneficent deity, and that alone was the key to the universe; all enquiry into the nature of the world was aimed, at least originally, at furthering an understanding of that relationship. It was a slow and sweepingly different set of approaches, framed by a need to prove on the one hand God's presence in everything; and on the other to perform a stocktake of everything that was known up until the 1500s which would ultimately refine and transform empirical methods of enquiry.

One of the notable exemplars of this enterprise was the Italian scholar Ulisse Aldrovandi (circa 1522–1605). In addition to being Professor of Natural History at Bologna University, he was also the Chair of Philosophy and Logic. Aldrovandi was determined to investigate what he called the secrets of nature as a means to better comprehending the workings of God. What made Aldrovandi a man way ahead of his time was his practice of conducting extensive fieldwork and engaging with locals to collect specimens, rather than relying solely on the received texts of antiquity. He was interested in the vernacular names of creatures brought to him, their habits and where they had been found. In addition, on his field excursions he took with him skilled draughtsmen – Lorenzo Bernini of Florence and Cornelius Swint of Frankfurt – to record whatever they found, and employed a secretary to take notes. While modern academics seldom have the luxury of secretaries in the field, they do rely on primary observation as a valuable source of empirical data. Aldrovandi was a pathfinder in this regard, practising both comparison and detailed observation. He taught his students to open up eggs to observe and understand what was happening inside them – the beginning of a new tradition of observation not limited by received knowledge as the boundary to enquiry.

Although most famous for his ornithology, dung beetles came into Aldrovandi's work in his 1602 fourth volume on the natural history of insects. He wrote that he found the study of insects the most difficult of all because of their size, and suggested that they should rather be called atoms. He speculated that it was because insects were so small that (as far as he knew) no one had published anything on them worthy of notice.

Aldrovandi's *De animalibus insectis libri septem, cum singulorum iconibus ad viuum expressis* (1602) provides further insights into his curious mix of modern observational science and traditional received wisdom. As a contemporary reader, it is often difficult to detect whether he is being critical of his predecessors, or is seeking credibility by quoting them; for example, he lists the various things that can give rise to dung beetles: the rotting flesh of horses (Isidorus), dead bullocks or asses (Virgil). Although it is not common, some dung beetles (such as species of *Scarabaeolus*) do use carrion as food instead of dung, while others vacillate between dung and flesh in their choice of filth in which to frolic. Even though Pliny the Elder (circa 23–79) correctly described the backwards progress of the ball-rolling beetle, Aldrovandi scolds him for not having 'accurate knowledge of *Scarabaeus pilularius*'. He proposes the common name *Scarabaeus* for dung beetles and promotes them as being '… splendid and distinguished. Even though they delight themselves in faeces they should not be an object of reproach, for even the alchemists use dung in their attempts to extract the fifth essence'[5] (part of the philosopher's stone)[6]. In colour, strength and in heroism, he ranks them above eagles.

In advocating for standardisation of common names for beetles and giving a clear description of *S. pilularius*, Aldrovandi might just take the honours from Linnaeus as the first scientific author (the person who first publishes the name of a new species) of a dung beetle species. This will remain debatable because Linnaeus developed the system which he himself used formally to describe all of the world's biota, including dung beetles. This makes it chronologically difficult to pre-empt him.

Nevertheless, the laurels for being the first to describe the nesting behaviour of dung beetles must go to our Italian 'father of natural history', as Linnaeus later described Aldrovandi. Preceding the French entomologist Jean-Henri Fabre by 300 years, Aldrovandi described and illustrated what is clearly the subterranean nest of a dung beetle, which he must have excavated in order to record what he found. His book is not only remarkable for its time in its attention to factual details along with accompanying illustrations, but also for its stocktaking of all that was known about insects at that point.

What hampered Aldrovandi was the lack of a framework within which to interpret his findings. His life's goal became to publish everything he could find on the known fauna of the world. The enterprise resulted in vastly expensive publications which few could afford, and cost him dearly. As an early collector of anything and everything, he displayed his acquisitions in a huge cabinet of curiosities, but the lack of a template other than a reference to Plato's four divisions of animals (to which he himself did not strictly adhere) meant that Aldrovandi's work teetered on a scale that oscillated uncomfortably between the past at one end and a distant future at the other, neither of which were available to him in a useful way.

Despite the upheavals of the century, Europe was criss-crossed by a variety of efficient postal services. Through this informal network, European scholars were able to share their ideas exhaustively; Erasmus complained that he had to write more than ten letters a day. As part of this 'republic of letters', Aldrovandi was in touch with another leading scholar in natural history who lived in Switzerland, Conrad Gesner (1516–1565). Like his correspondent Aldrovandi, Conrad Gesner was attempting to gather all the known records of the world's flora and fauna into a series of books. Together these two encyclopaedists represent early examples of a foundational aspect of research: gathering all previous known work in a particular field of enquiry as a starting point for further research. Gesner gathered as many accounts of animals as he could find (as well as drawings from different sources) which he then recorded and attributed. While his work included mythical animals, its real significance was the systematic recording, ordering and annotation of information from diverse sources. Although in his descriptions of plants he used the taxonomic categories of species, genus, order and class that we recognise today, Gesner didn't progress beyond Aristotle in his perceptions of the patterns of biological order.

Like his contemporary Aldrovandi, Gesner was however laying a foundation for the later empirical tradition; however, without a template or broad interpretation of the huge subject matter he was dealing with, his work was primarily a record rather than a reframing of information. Even his tentative understanding of the concepts of genus and species was not pushed to its logical conclusion and the fame of providing a framework to understand nature passed this remarkable man by.

Aldrovandi's studies found their way into a work Gesner began but did not complete, and which was to become a scientific game of pass-the-parcel in its path to eventual publication. Gesner had begun writing a book on insects but had not completed it at the time of his death during the plague in 1565. The manuscript was then sold to the theologian and physician Thomas Penny (circa 1530–1589) who had visited Gesner. Penny proceeded to combine it with notes and observations of the Oxford physician, Edward Wotton (1492–1555), but the book remained unpublished. After Penny's death the manuscript passed on to his neighbour and friend, Thomas Moffet (1553–1604), who completed the work and even commissioned its title page. The manuscript nevertheless did not see the light of day until 30 years after Moffet's death when Sir Théodore Mayerne (1573–1655), a successful physician, finally published the book *Insectorum sive Minimorum Animalium Theatrum*, also known as *The Theatre of Insects*, in 1634.

Moffet's title page displays a handsome ball-rolling dung beetle. The inspiration for this illustration is one of conjecture, as there are no ball rollers in England's limited dung beetle fauna. Moffet had travelled to both Italy and Switzerland, but he was not the artist. It is likely that some of the illustrations and information used in the *Insectorum* were based on the work of Aldrovandi's draughtsmen, in an era when it was common practice for art to be transferred and copied repeatedly, particularly when woodcuts were used as they were in *The Theatre of Insects*.

Moffet's book is regarded as the first English-language book on insects, but it owes its existence largely to the tradition already established by Aldrovandi and Gesner. It represents a very distinctive strand in the early practice of natural science, one that was not that successful in countering a less empirical approach. It is clear that the author understood that dung beetles were revered by the Egyptians and it repeated the Egyptian belief that there were no female dung beetles. It is an intriguing muddle of empirical observation, speculation and traditional wisdom. On the one hand, it speculates that farmers might well have learnt to compost their fields based on their observations of the effect of dung beetles distributing their balls of dung; on the other it notes that Saturn is the dunghill god and offers the not uncommon moralistic comparison between humans and dung beetles, that it is the dung beetle which stays relatively clean and neat in dung, whereas humans are polluted and infected with flaws. Although observation-based knowledge was becoming more systematic, the persistence of magical and mystical beliefs in publications such as *The Theatre of Insects* points to the continued conflation of the empirical and the magical.

It was a road that was to survive well into the next century, and one which even Isaac Newton (1643–1727), often described as the first of the modern scientists, was to travel throughout his life. Newton wrote more about alchemy and magic than he did about physics. John Maynard Keynes described Newton as 'the last of the magicians, the last of the Babylonians and Sumerians ... the last wonder child to whom the Magi could do sincere and appropriate homage'.[7] So it is not surprising that Aldrovandi and Gesner included mythical creatures in their encyclopaedic works, which they wished to be as comprehensive as possible. Nevertheless, most of what they reported was rooted in the observable and the recorded. It was not based on alchemical practises and beliefs that pervaded much of the emergent scientific thinking of the time. This seam of thought owed as much to the rediscovery of ancient writers during the Renaissance, which had included the emergence of old hermetic texts, as it did to the centuries of alchemical and magical ideas that preceded it. Unsurprisingly, Khepri reappeared somewhat divested of his former splendour in these arcane records.

Hermeticism is based upon the writings of the ancient (probably mythical) prophet Hermes Trismegistus (who is credited with being a contemporary of Moses). These were mystical texts that pursued fundamental questions about the ingredients of life, described as the original truth. Egyptian religion (as conveyed through Hermeticism) emphasised knowledge, which came about both through revelation (the understanding of symbols), and through the power of magic (as exhibited through alchemy and divine protection). Ancient gods and goddesses symbolising different energies and powers persisted via Hermetic philosophy, and even formed part of the Christian discourse of the medieval period.[8] By linking the Renaissance directly back to ancient Egypt by association with the Egyptian priest Imhotep (2667 BCE–2600 BCE), who was proposed as one of the texts' authors, the sacred scarab may well have worked its way back into the limelight via its original promoters.[9]

One man who brought these divergent strands of religions, hermeticism and dung beetles together into the heart of the Catholic Church was Athanasius Kircher (circa 1601–1680). Born in Germany, he became a Jesuit at the same time that the Thirty Years' War began.

He travelled widely and ultimately ended up in Rome, where he befriended the high and mighty, including Pope Urban VIII. Kirchner learned Hebrew as a young man, and came to believe that the solution to all communication problems was the development of a common language. He believed that Egyptian hieroglyphs were the magical key linking symbolic thought with language, and that once such thought was unlocked, humankind would be brought closer to God.

In pursuit of his theory, he set out to translate and understand hieroglyphs, basing his research on the works of Horapollo, Plutarch and the Mensa Isiaca,[10] a famous Roman table reputed to be the table of Isis, but which turned out to be a fake with meaningless hieroglyphs. Kircher devoted much of his time to his great opus, and finally published his *Oedipus Aegyptiacus* between 1652 and 1654. The work was a failure, but is of interest to us as he included a curious and fascinating image of a scarab rolling a ball of dung through the planetary spheres.

It is likely that Kircher would have come across the scarab in the Coptic writings that he had access to via the French astronomer, savant and antiquarian Nicolas-Claude Fabri de Peiresc (1580–1637). Kircher had many strange and fascinating ideas, not least of which was that all water was stored in a massive subterranean sphere under the earth, with rivers springing from this source. In a short manuscript he wrote entitled *Mundus Subterraneus* he suggested that awareness of what he termed 'panspermia'[11] was first developed in ancient Egypt, and it was the dung beetle with its ball 'orbiting the globe, [which] gives it life and fertility by the means of the same panspermia rerum and fills it with every kind of thing.'[12] For Kircher, the scarab provided a vivid demonstration of the belief that putrefaction bred life.

It was a widespread conviction that accorded well with the concept of the spontaneous self-generation of insects, a belief inherited from the writings of both Aristotle and Pliny. There can be no doubt that Kircher understood some of the significance of Khepri to the Egyptians, but his inaccurate understanding of the hieroglyphs meant that he did not place Khepri properly in Egyptian cosmology, and instead introduced him in a modified and altered form into the alchemical realm he found so compelling. Kircher's writings were immensely influential in the undefined territory of the time between chemistry and alchemy; they captured the reigning current of confusion between knowledge that threatened traditional Christianity alongside concepts of sun gods orbiting globes, which he seriously believed entertained varieties of acceptable truth about the workings of the physical world.

Kircher's fanciful muddle did not appear to earn him any significant opprobrium from the Catholic Church. Galileo Galilei's assertion that earth was not the centre of the universe was a challenge of an entirely different order. Kircher's patron, Pope Urban VIII, who left Kircher in peace to write and publish his strange theories, was to imprison Galileo (1564–1642) for his radical discoveries, which the Inquisition denounced as a direct challenge to the Church. Galileo was one of the first men to openly undermine the received wisdom of Aristotle, and to question the Church. His support of the Copernican theory of a sun-centred solar system led to him being accused of heresy. He was found guilty and placed under house arrest until his death. It seems incomprehensible that what is now so obvious represented such a challenge to the power of the Church. History is inevitably littered with the detritus of such discarded beliefs, but the process is rarely rushed, linear or clear, as supported by the curious fact that the Catholic Church only officially reversed their decision on Galileo's 'heresy' in 1992, while Kircher continues to be somewhat dubiously described as 'the last man who knew everything'.

In Northern Europe, a contemporary of Galileo's, Rene Descartes (1596–1650), was preparing an equally fundamental challenge to the Church. His experience of the Thirty Years' War had led him to a profound intellectual response to the consequences of religious uncertainty: a new philosophy that began with the simple but startling premise, 'I think therefore I am.' This evolved into four steps that ultimately underpinned the development of the modern empirical scientific method. The first step was to accept nothing that could not be clearly recognised to be true. The second step was to divide up each of the difficulties into as many parts as possible so as to resolve the question in the best way possible. The third part was to start with the simplest and easiest issue in order to rise to the most complex, even if the order was a fictitious one, and there was no natural sequence. The last point was to make notes as complete as possible and compile reviews in which none of the methods used were omitted. Such deceptively simple order and logic posed part of the scientific challenge both Descartes and Galileo introduced into a world still inclined to entertain the old magical symbolic thinking of Kircher. It would take time to dislodge the latter, but once the door had been opened, it was impossible to close it. The tension between the two opposing worldviews nevertheless persisted even in progressive environments, where empirical observation and methods were beginning to reverse centuries of traditional practice.

Dutch medical schools of the time had many outstanding scholars who straddled the two worlds while introducing new medical practices. Despite personal interests in the arcane, the progressive and innovative changes of teachers such as Otto Heurnius were to cement the growing reputation of Holland as a centre for medical excellence. Attracted by this repute Carl Linnaeus (1707–1778), a Swedish medical student, chose to complete his studies in Holland and received his degree of Doctor of Medicine from Harderwijk University. He then enrolled for further studies at Leiden University. Holland was experiencing what is now called its Golden Age, stimulated in part by the ending of Spanish rule, as well as its control over resources from the Far East via the Dutch East India Company or Vereenigde Oostindische Compagnie (VOC). The VOC, a key player in the resurgent strength and wealth of Holland, was founded in 1602. It was to become the first multinational company in the world, and evolved a number of strategies to maintain its monopolies. One of these was the establishment of a tight grip on its holdings, including forbidding employees from disseminating knowledge about anything found within territories under the control of the VOC. The only reason that objects with no obvious commercial potential arrived in Holland was because of the demands made by individual directors, rather than any scientific interest or desire on the part of the company to learn about anything other than profit.

Most of the directors of the VOC demonstrated a mercantile mind-set, but there were exceptions, and George Clifford III (1685–1760) was one of them. It was at his famous country estate Hartekamp near Harlem, lavishly endowed with specimens brought back from exotic destinations, that Linnaeus had eagerly sought employment after encountering what he ecstatically called the four houses of Adonis. These were hothouses filled with the flora of Africa, Europe, Asia and the Americas, as well as a museum of which he declared that he had never seen the like of before.

Linnaeus had demonstrated an interest in the naming of plants from a young age, and the combination of his work back in Swedish Lapland, and his studies in Holland resulted in his remarkable and seminal publication *Systema Naturae* (1735), a catalogue of all the species of the world. Although initially only 12 pages long, it was the equivalent of splitting the atom in the way it was to transform how all information about the flora and fauna of the world was to be organised.

He used flower structures to order them into related groups, and described the sexual parts of the flowers in Latin to reveal how species that might appear very different may in fact be closely related. Despite the immediate success of this logical system, Linnaeus was criticised as lascivious because of his emphasis on sex. The *Polyandria*,[13] for instance, translates as 'twenty or more males in bed with the same female'.[14] The Bishop of Carlisle said that 'nothing could equal the gross prurience of Linnaeus's mind.'[15] Needless to say, despite the inevitable criticisms of eighteenth-century trolls, the system was a success and forms the basis of the binomial nomenclature we still use to name all living organisms.

However, this was not simply a way of ordering information; it also represented something more profound to scientific thinking of the day. A hundred years earlier, Descartes in his 1637 *Discourse on Method*, had suggested that even if there was no order in a subject, it would be necessary to create an artificial order as part of the process of coming to know the likely answer to a question. Linnaeus had successfully taken up that baton and presented the world with what became an enduring structure for ordering information.

In the world of dung beetles, Linnaeus began to unravel the confusion surrounding the naming of what until then had loosely been defined as scarabs. He started appropriately by formally describing the sacred scarab (*Scarabaeus sacer*) within his own binomial system in 1758, in the tenth edition of his *Systema Naturae*. Specimen 3344 can still be found in the collections of the Linnaean Society, with a locality label bearing the script 'Algiers' attesting to its North African origin. It is an archetypal, big, ball-rolling dung beetle with the characteristically large eyes of a nocturnal species. This leads to a rather intriguing question: if our iconic Egyptian beetle is nocturnally active, how then did the sacred scarab earn its reputation of being associated with the movement of the sun?

The waters are further muddied by the genus name *Scarabaeus* being attributed in 1810 to *Scarabaeus hercules*, by Pierre Latreille (1762–1833). Latreille was an orphan who became a Catholic priest and renowned French entomologist. His knowledge of beetles led to his being released from a revolutionary jail when the visiting doctor found him engrossed in a rare specimen found on the floor of his cell. Described as the 'prince of entomologists,' Latreille introduced the concept of a 'type species'. This concept is one in which an actual physical specimen is designated the 'holotype' of a particular species, making it the permanent representative of that species, usually available in museum collections for examination by experts and now often online as a series of high-resolution photographs. Importantly for our story, the genus name is usually also attached to the type specimen. But somehow Latreille gave that honour not to a dung beetle, but a rhinoceros beetle, *Scarabaeus hercules* (the eight-centimetre long monster now classified in the Dynastinae, a different subfamily altogether to the Scarabaeinae and which is now usually called *Dynastes hercules*). Fortunately for the romance of the name 'the sacred scarab', the strict rules of zoological nomenclature are sometimes ignored, and *Scarabaeus* has remained with the dung beetles, but the tangled web of names around a single, well-known if not actually famous beetle illustrates the difficulties we still face in trying to classify organisms into logical groups. If one imagines the difficulty, for example, of trying to pick a single holotype to represent the glorious variety found in all the shapes and sizes of *Homo sapiens*, then the problem of doing the same in the infinitely more varied world of beetles becomes clear.

The naming problem of the sixteenth and seventeenth centuries was exacerbated by the sheer volume of unnamed flora and fauna being introduced to the West through blossoming trade routes. Wealthy Dutch collectors were avidly filling cabinets of curiosities known as *kunstkammers*. These formed condensed visual representations of a radically new world view, and contained all manner of objects, unfortunately not usually well-preserved. Nevertheless, some of the contents of these early collections can be deduced from paintings of the plants and animals by notable artists of the time, many of whom were keen naturalists. In the absence of photography, art was to become the realm in which some of the earliest depictions of the new discoveries in the natural world were recorded. In the process the artists became some of the earliest exponents of the empirical tradition by relying on evidence and observation to construct their paintings.

One of our first sightings of a seventeenth-century depiction of a dung beetle is in the painting by Pieter Holsteyn (1585–1662) of a rhinoceros beetle and six other insects, one of which (the black and yellow beetle) might be a fruit chafer, a related scarab. The rhinoceros beetles are in the subfamily *Dynastinae* and the fruit chafers in the subfamily *Cetoniinae* (which along with the *Scarabaeinae*, the true dung beetles) are all members of the family *Scarabaeidae.* These delicate renditions of exotic insects are valuable not only for their beauty and accuracy, but because they provide an insight into changes in sixteenth-century concepts of nature. Holsteyn was himself following in the footsteps of the Flemish Hoefnagel family of artists, who had depicted the intimate and tiny world of the enormous variety in insects in unique and novel ways.

The Hoefnagel duo of father and son, Joris (1542–1601) and his son Jacob (1575–1630), were pivotal in establishing a tradition of illustrating insects and nature as more than minor decorative details in grand paintings. Joris in particular enriched entomology with his accurate and beautifully rendered paintings of insects. He was well educated and widely travelled, as was Jacob, who followed in his father's footsteps, dazzling the imperial court in Prague where he was court painter to Emperor Rudolph II from 1602 onwards. While still a teenager, he worked under the supervision of his father on the *Archetypa studiaque patris Georgii Hoefnagelii*, based on his father's designs. It was published in Frankfurt in 1592 and contained 48 engravings of plants, insects and small animals, with each of the plates arranged as if the specimens were items from a cabinet of natural curiosities. The work can be viewed almost as a bridge between the old world (with its use of epigrams from Erasmus's *Adagia* and its focus on religious symbolism as a means of representing God in his smallest creatures) and a new world where insects were finally being accurately depicted and seen as objects of scrutiny in their own right.

Earlier artists, including Albrecht Dürer (1471–1528) had indeed painted or drawn insects accurately, but the insects in those paintings were used symbolically (most notably the stag beetle, which was used as a symbol of Christ).[16] Insects that had no symbolic value seldom if ever featured in paintings, but the new depiction of insects divested of any such role created links between artists and naturalists. Joris Hoefnagel, for example, was known to have relied on Conrad Gesner's work for some of his sources. It was a numerically small but busy artistic and scientific world that mirrored the little creatures on which they were all focusing.

Despite the extensive trade networks of the Dutch and the thirst for new information, there were significant gaps and oversights in the collection of important publications from distant countries. A fine example of this was the *Bencao Gangmu* (1578). Compiled by Li Shizhen in China, it was contemporaneous with the Hoefnagel publication, but this authoritative and comprehensive Chinese *Materia Medica* had illustrations that were nothing like the delicate renditions of Joris's publication. It is hard to quantify or describe the huge chasm between the West and the East at this time. The VOC was trading with the East, and had its employees actively recording and appropriating indigenous knowledge, but the focus was exclusively on what could be usefully profitable in the West. Amongst the small group of artists painting the flora and fauna presented to them, it is only in the inclusion of objects from the East (such as beautiful china) that we can observe any presence of what must have seemed a very distant and exotic culture.

Artists recording the new flora and fauna, even though depicting their subjects with remarkable accuracy, did not exclude their religious beliefs from their paintings. The Hoefnagels firmly believed that it was God's abundance they were illustrating, but their paintings were nevertheless shifting the traditional use of insects in art from a didactic to a purely decorative role. Early seventeenth-century Dutch art represented a visual world dense with meaning. Everything in a painting conveyed a message, from everyday objects depicted (a pipe was sexual, for instance) to a gesture such as a letter being read (meaning that a lover or spouse was abroad): an entire story was communicated to the viewer. Insects were not simply insects; they were representatives of qualities both good and bad, which was one reason the array of insects in earlier paintings was so limited. The introduction of new insects with no known metaphorical role in the Hoefnagels' ground-breaking paintings represented a small seismic tremor in perception.

The tremor became more of an earthquake in the world of dung beetles as the impact of the single lens microscope, used by Johannes

(Jan) Swammerdam (1637–1680), enabled him to both describe and draw a scarab and its immature stages, as well as to draw the larva and pupa in his *Bybel der Natuure*, published posthumously in 1737. It is a remarkable work, in which the scarabs chosen are once more rhinoceros beetles, probably because the European species are large enough at all life stages to be seen clearly even without a microscope, and their larvae are easy to rear on compost or manure.

Swammerdam was a complex man who encapsulated the spirit of the times with his constant search for answers in nature that satisfied both his faith and insatiable curiosity. Trained as a doctor at Leiden University, his lifelong passion for insects led him to focus on them rather than human patients. Inspired by Marcello Malpighi's monograph *De Bombayce* (1669), a pioneering account of the microscopic dissection of silkworm larvae, he began to combine his dissection skills with the use of a microscope. After publishing his *Historium generalis insectorium* in 1669, in which he debunked the notion of the spontaneous generation of insects, he went through a profound spiritual crisis. Although Cosimo de Medici (to whom Swammerdam had demonstrated that the future butterfly could be detected in the caterpillar) had offered him a position at the Medici court in Florence, he instead left Amsterdam to join a small community that had formed around the mystic Antoinette Bourignon at Nordstrand, an island off the coast of Schleswig-Holstein.

Bourignon had told Swammerdam that his work on the silkworm (which he subsequently burned on her instructions) encouraged Satan, but she quixotically gave him permission to publish his research on the mayfly prior to his setting out to join her community. He was immensely troubled at this time by the conflict between his pursuit of science and his religious ideals, and most of his treatise on the mayfly was devoted to a meditation on the idleness and sinfulness of science. He eventually came to believe that he needed to turn away from the forbidden tree of science.[17] This fear of an incompatibility between science and his beliefs ultimately resulted in a fortuitous outcome for the early days of entomology: having become disillusioned by Bourignon, Swammerdam returned home from Nordstrand and spent the next four years of his life pioneering a microscopic study of the anatomy of insects. Like an earlier and much neglected fellow naturalist, Johannes Goedart (1617–1668), Swammerdam had come to believe that if God had in fact created everything on earth, then everything was worthy of human attention. Life's complexity could reveal the handwork of God, and science was useful because it could unravel the mysteries of God. For the insects of the world this was superb news, as this meant that they became acceptable subjects of serious study and interest.

Swammerdam dissected a stag beetle in 1673 and was astonished by its genitalia: 'consisting of no more than a single long, hollow, innumerably inflected fine thread'.[18] He presumably had focused on something other than the *aedeagus* (the insect equivalent of a penis), which in beetles and the scarabs in particular is an impressive chitinised (rigid) structure usually about a quarter of the male's body length. It is possible that he might have dissected a female, but because they lack the massive mandibles of the males, it is all but impossible to mistake one sex for the other. Despite this confusing detail, the giant step Swammerdam's work represented was only possible because of the invention of the microscope. Although Swammerdam used a single lens microscope based on a very simple design, his contemporary Antonie van Leeuwenhoek (1632–1723) is now credited as being the first person to make a real microscope.

This was a piece of technology as significant as Linnaeus' publication. The world was acquiring a new understanding of structure; what had been small and invisible to the naked eye was becoming visible. This development paralleled the telescope's revelations of planets and moons, and in some ways the discovery of continents and peoples on the other side of the globe. These discoveries transformed everything. The scale of life itself changed: the world was much larger than had been imagined, and so was the cosmos; and this applied equally to the microscopic world of insects that was finally being revealed.

The world exposed by the use of a microscope transformed the understanding of insect anatomy, but it wasn't until the German-born Maria Sibylla Merian (1647-1717) published her paintings that the complex relationship between insects and their natural environment and predators was recorded. Merian was the stepdaughter of the artist Jacob Marrel (1613–1681), noted for his paintings of the tulips so beloved of and damaging to the Dutch economy. Merian's paintings were to introduce what we now call ecology: a huge although logical step from spotlighting individual insects. Her achievement was all the more notable because she made a living as an artist at a time when women artists were few and far between.

Her reputation as an artist was well established in Holland, where she lived for most of her life. However, her desire to learn more about insects and plants led her to travel to Surinam with her daughter Dorothea Maria, in order to observe first-hand the world she had found so beguiling by proxy, through the specimens she was given to record. Merian was one of the first explorers to acknowledge indigenous informants as valuable sources. She often relied on them to bring her specimens for what became her magnificent set of paintings of the flora of Surinam, which showed the symbiotic relationships between plants and their herbivores. By breaking the tradition of a separation of species and their natural environment, she established a new direction in the depiction of nature. Her paintings were so accurate and detailed that Linnaeus and his students were able to classify over one hundred species based on their study of her illustrations.

Her work was to have a lasting impact on August Johann Rösel von Rosenhof (1705–1759), an Austrian aristocrat who had the good fortune to be educated as an artist after his godmother noted his natural predilection for painting. Rosenhof began his career painting miniatures and portraits at the Danish court in 1726. Returning home after two years in Copenhagen he fell ill, and during his long convalescence he was given a copy of Merian's book. The work was a revelation to Rosenhof, and inspired him to research, write and illustrate a book on German fauna. Like Merian, he collected many of his subjects and raised them in his home in order to observe and record their development and metamorphosis. The result was two large books. The first, *Insecten-Belustigung* (Insect Amusements), appeared in 1740.[19] In this volume, he classified insects in natural groups, and this later earned him the reputation of being the father of German entomology. His paintings of scarabs are vivid, but mostly illustrate rhinoceros beetles. As with Merian's work, Linnaeus used many of Rosenhof's illustrations and descriptions of species as the basis for naming them within his own classification system.

Another artist and naturalist who was influenced by Merian's ecological approach was the naturalist and explorer, Mark Catesby (1683–1749), who travelled to America twice in the early eighteenth century. Catesby's records were so valuable to Linnaeus that they formed the basis of his formal descriptions of American species. On Catesby's second journey, begun in 1722, he spent four years recording the flora and fauna of South-Eastern North America and the Caribbean. During this period, he became the first person to paint and record in situ an indigenous North American dung beetle. His painting of two 'tumble turds' (as he called dung beetles) contains two dung beetles busy around the base of a *Lilium* sp *Martagon canadense*. One of them, portrayed with a characteristic ball of dung, is *Canthon pilularius*, a common North American species, while the other beetle heading off the right-hand side of the page is *Phanaeus vindex*, the rainbow scarab beetle, a very pretty shiny green tunnelling species.

'Tumble turd' seems to have been a common name for dung beetles in North America; not long after Catesby's publication, another account of them appeared in *The Natural History of North Carolina* by John Brickell, published in Dublin in 1737. This book included a diverting account of the tumble turds, 'so called from their constant rowling (rolling) the horse-dung (whereon they feed) from one place to another till it is no bigger than a small bullet. They are one of the strongest insects of the same size I have ever seen; they frequently fly into houses, and I have seen one of them move a brass candlestick from one place to another upon a table which seemed to me very strange at first.'[20] Evidently Brickell's host placed two tumble turds underneath two candlesticks and made some supposedly magical noise that caused the beetles to start their Herculean efforts to escape their brass prisons; it was this trick that occasioned the comment. He noted that the beetles seemed to be larger than their relatives in Ireland, and described them as being infected with small light-brownish insects, commonly called beetle lice. These passengers are phoretic (hitchhiking) mites, which (being flightless) use the dung beetles for passage between relatively widely spaced dung pats where they prey on other invertebrates. The mites have entered dung beetle lore as potential biocontrol agents of dung-breeding flies in Australia, a role they have failed to fulfil. A further erroneous theory about these mites was that they lived symbiotically with dung beetles and helped them to find dung. Brickell didn't do much better, claiming that the mites' 'powder is used against the falling out of the Fundament, to expel urine and cure the bite of a mad dog. The juice cures wounds and in plasters buboes and pestilential carbuncles'.[21]

Anecdotal accounts by somewhat dubious writers like Brickell were entertaining for readers, but the new spirit of eighteenth century enlightenment was abroad, and a different calibre of field research was developing. Pehr (Peter) Kalm (1716–1779), Professor of Economy at Abo University in Sweden, and one of Linnaeus's earliest students, represented this change during his travels in North America scouting for novel plants that might be useful back home. His observations of dung beetles were more detailed and detached than those of his predecessors. In May 1749, he noted that 'The dung beetles had dug very deep into the ground, thro' horse dung, tho' it lay on the hardest road, so that a great heap of earth lay near it. The holes were afterwards occupied by other insects, especially grasshoppers, and cicadae.'[22] Kalm was not preoccupied with proving God's hand in the making, or changing, of America. He was intent on observing and recording, and in so doing, he represented a fundamentally altered approach to the way in which nature was chronicled.

Kalm's work was a measure of the distance dung beetles had travelled: from being infused in oil for remedies for the poor in the Middle Ages to annotated, ordered and beautifully rendered paintings of them in the seventeenth century. During this long period, traces of dung beetles were initially rather faint; but via a circuitous path (with tributaries incorporating metaphor and hermetic imagination) they emerged via art and ordering into the clearer light of observation and description. By the end of the seventeenth century, some of the foundations of modern scientific entomology had been laid, but these were by no means more than a base. A huge number of specimens remained unknown, and the massive enterprise of discovering and naming them was one that would accelerate in subsequent centuries, leading to even more profound changes in our understanding of the natural world.

CHAPTER THREE
Joining the dots

IN THE LILLIPUTIAN WORLD OF INSECTS, the seventeenth century was one of seismic change. They were no longer ignored or seen mainly as symbols for everything from industry to sinfulness; instead they had found their way into starring roles in contexts ranging from display cabinets in the drawing rooms of the wealthy, to learned books and paintings. Like so many of the places seen as newly discovered, insects had always been there; what had changed was the nature of human awareness and perception of them.

In territories across the globe they were still sources of food and medicine and featured in local narratives and art, but the great enterprise of collecting and organising the information being assembled by travellers and explorers was reducing much of that historical relationship to passing references or anecdotal information. The Linnaean system of naming flora and fauna was gaining traction among collectors, and the desire to own new and unidentified specimens was helping to fuel the expansion of private collections. However, it is one thing to start collecting beetles or shells (or indeed anything that catches the eye) but as anybody who has walked on a beach and collected a bag of shells knows, they don't have the same allure once removed from the shore, unless they are displayed or used for some purpose.

Collectors faced a number of problems, not the least of which was what to do with their collections after their death. Having spent years assembling unique collections of exotic material, often at great cost, the thought of everything being disbanded or (worse still) disposed of by disinterested heirs was not to be countenanced.

Sir Hans Sloane (1660–1753), one of the great collectors of his age, found a profitable and enduring solution by bequeathing his very substantial collection of 71 000 objects to King George II 'for the nation' on condition of a payment of £20 000 to his heirs.

Sloane was a medical doctor who had dedicated his book on the natural history of Jamaica to Queen Anne, describing the 11 000km² island as 'the largest and most considerable of Her Majesty's plantations in America.' Implicit in this is a world in which monarchs owned entire countries, and an island was merely a plantation as opposed to a sovereign territory. This dedication captured the essence of empire, and the values of the eighteenth century.

Sloane's collection became the core of the British Museum, founded in 1753. It was the first museum of its kind in the Western world, and eventually the home of some of the largest and most valuable collections of cultural artefacts of the globe. Sloane himself demonstrated a passion for naming and claiming that was to typify the rise of the British Empire. This practice was critical to the important scientific questions and advances that would follow. The eighteenth century was to see the transition from private to public collections, with natural history emerging in its modern form as a scientific subject.[1]

Kunstkammers had largely been dictated by the private passions and wealth of a few enthusiastic individuals who gathered the first fruits of global exploration in their cabinets and homes. Peter the Great, Tsar of Russia (1672–1725), was a prominent collector and visited Western countries in pursuit of rarities that became the basis of his collections. He predated the British Museum with his establishment in 1727 of the *Kunstkamera* in St Petersburg; its strange contents (which included many deformed bodies) placed it firmly within the ranks of earlier *kunstkammers,* which reflected personal predilections rather than scientific pursuit. Many of the Dutch *kunstkammer* collections were sold to him, giving Russia an impressive assemblage of some of the world's most exotic finds – until the building they were housed in burned down. Only a fraction of the specimens were saved.

Strictly speaking, France's Natural History Museum is the world's oldest natural history museum, having been founded by Louis XIII in 1635 as the Jardin Royal des Plantes Medicinales. In 1793, following the French Revolution, it became the Museum National d'Histoire Naturelle and employed many notable French scientists.

These early museums served the larger enterprise devoted to acquiring and naming the flora and fauna of the planet, in an intimate, almost nepotistic world where money and lineage helped primarily because those able to pursue their interests in natural history were usually individuals of independent means.

There were some exceptions to this among the clergy and academia. It was mostly a convivial world (with some notable lapses), but letters written between collectors usually described their acquisitions with much enthusiasm and in great detail. Everything was recorded: from the first flowering of a new plant, to the mechanics of installing a thirty-foot stove to keep flower beds containing rare and exotic specimens warm. Linnaeus presided over this world, and was regarded with great respect; for collectors such as Francis Masson (1741–1805), a humble plant and insect collector in Southern Africa, having Linnaeus name a plant genus (*Massonia*) after him represented the pinnacle of his career.

William Kirby (1759–1850), a parson and entomologist who named many members of the family Scarabaeidae, exemplified the sedate inter-connectedness of the world of entomology in the eighteenth century. Kirby graduated from Cambridge in 1781 and having no expectation of any inheritance as the son of a topographer, took holy orders instead in 1782 and became the parson of Barham in Suffolk, which position he held for 68 years. Although a relatively obscure figure, he was one of the many individuals who, in their devotion to their collections, paved the way for entomology to emerge as a science. He became a Fellow of the Linnaean Society, joining other enthusiasts who enjoyed the hunt for specimens to enlarge their own collections. Always looking for new material, they stimulated a network of international collectors who slowly and methodically established taxonomic order across the vast diversity of the insect world. They were accountants of nature, adding, subtracting and ordering information. They were not adventurers, and seldom displayed any boldness of vision, but these accountants helped make sense of the dizzying variety of specimens being brought to Europe from all over the world.

The individuals going out and gathering flora and fauna were quite different from the collectors who created the demand for exotic material. Intercontinental travel throughout the eighteenth century was difficult and uncommon. It became a way for adventurous travellers to fund their travels and to make a decent living, with many collectors willing to pay handsomely for sought-after items. Dru Drury (1724–1804), an enthusiastic entomologist who in a letter to a collector charmingly declared that 'Insects are my darling pursuit', published the first book in English of exotic entomology, *Illustrations of Natural History*, using the Linnaean binomial system. [2]

Drury is of particular interest to the history of dung beetles because of his focus on them. He had a notable collection, accumulated over 30 years of commissioning, begging and badgering individuals travelling in remote countries, as well as inhabitants of those countries. His collection, comprised of far more than dung beetles, cost him £4 000 to acquire and a further £1 000 to maintain. He offered a gardener in Jamaica sixpence apiece for any insects he gathered, and was then concerned that it might not be enough. Considering that most insects in gardens were probably eating something usually considered more precious than the insect itself, men like Drury might well have been an early source of the 'mad dogs and Englishmen' concept. He was not indifferent to such matters, and in a letter to a Mr Smeathman who was travelling in Sierra Leone in 1772 asked whether the 'Blacks relish your catching Birds and Flies, whether they laugh at you for so doing.'[3]

Drury was obsessed with acquiring specimens. He took any and every opportunity offered by ships sailing abroad to encourage travellers to gather material on his behalf, and was not above requesting individuals on board slavers to collect for him in West Africa. He merely bemoaned the avarice of the merchants for 'carrying their ships such an enormous way around as ye West Indies and not sending them directly to Europe.'[4] A typical stay-at-home collector of his era, he was a 'gentleman' whose excitement over nature knew no boundaries, and whose interest in humans was clearly minimal. He exhorted a friend: 'not let ye summer pass without making captive all ye insects that fall in ye way.'[5]

Drury wished that commissioned collectors would make notes on the environment in which they found their insects, but he knew that the chances of this were small so he settled on the tasks of collection and identification. He was in many senses no different from his beloved insects – a product of his time and environment. He was, however, an indication of what was to come because he sensed that understanding fauna within their environments would possibly provide answers to the puzzles raised by the variations and differences in the species being collected. His collection was substantial and when it was sold in 1805, items such as his *Scarabaeus goliath* were bought by G. Humphrey for £12.1.6.; while 13 specimens of *Bupestris* (spectacular metallic 'jewel' beetles) fetched £8.0.0.

The rise of the British Empire fuelled the acquisition of artefacts and collections for the British Museum, but such enthusiasm was not confined to the British. French scientists were equally prominent in the late eighteenth century. The French were as keen as their neighbours, the Dutch and the English, to explore the implications of a world of so much variety. They were among the first to demote the concept of a great chain of being, the kind of hierarchy in nature that the Dutch were untangling in their paintings. For beetles, the demise of the notion of the chain of being was critical to their adoption into the world of nature as more than simply creatures at the bottom of the natural order. They were moving up in the world – but even though the great chain of being was disappearing, hierarchy was not.

Hierarchy, or the need to establish structured order, was to some extent, an inevitable outcome of the continuing collection and naming of global flora and fauna. Linnaeus (and subsequently his student, Fabricius) were two significant players in this particular exercise, and the museums and natural history collections being formed were an intrinsic part of the process. However, while identification was the starting point, an even more intriguing set of questions arose based on the obvious similarities and differences within the many species being assembled. It was clear that there were relationships and families scattered all over the world (rather like us humans), and questions about their histories and differences were beginning to absorb naturalists.

Dr William Hunter (1718–1783), a keen and discerning collector whose collection was to form the basis of Scotland's first museum (The Hunterian in Glasgow), gave an indication of this line of thought when he commissioned George Stubbs (1724–1806), the famous English painter of thoroughbred horses, to paint a young bull moose. Hunter was curious to see whether the North American moose was the same species as the European elk. As a medical man with a particular interest in anatomy, Hunter's attempt to answer questions about species started with comparisons of anatomy. His collections were not confined to creatures with skeletons: amongst his large collections he had an insect collection in need of identification and organisation, which Johan Christian Fabricius was invited to curate.

Fabricius (1745–1808) is one of the most significant people in the eighteenth-century world of dung beetles because he named substantial numbers of them, becoming the 'author' of these species. He identified an impressive 9 776 insect species in his lifetime, and was affectionately christened the Linnaeus of the insect world. Fabricius was well loved, and he and his fellow students reportedly danced in the early mornings on the lawn with their teacher Linnaeus. Unlike the latter who grew wealthy and pompous as his reputation grew, Fabricius never earned much money in his lifetime, and remained a humble and popular character. He was appointed Professor of Natural History and Economics at the University of Kiel, a position he held throughout his life. Despite this employment his income was small, but this proved to be a boon because it stimulated Fabricius to travel in order to curate and identify specimens for wealthy collectors. In 1767 he went to Britain for the first time, where he met a fellow former student of Linnaeus, Daniel Solander (1733–1782), then working as a cataloguer at the British Museum.

Solander introduced Fabricius to the small circle of naturalists and enthusiasts that included Joseph Banks, Dr William Hunter and Dru Drury. Fabricius was then employed to curate and name Drury's collections of insects, as well as other collections; and so the earliest identification and naming of many species of the *Scarabaeidae* came about. The significance of this went far beyond a simple act of naming, as Fabricius laid the foundations of modern taxonomy and entomology. The English embraced the Linnaean system with a fervour that bordered on the religious. Fabricius himself noted that 'the number of species in entomology is almost infinite and if they are not brought in order entomology will always be in chaos.'[6]

This is clearly true: without accurate identification and description, an insect simply remains an insect. The difference, however, between giving a scarab a species name (as opposed to the status of a god based on imagined similes) lies in finding what qualities and characteristics define that scarab from the rest of its close relatives. The word *Scarabaeus* itself came from two Greek words, signifying 'to walk on the head'. This is clearly based on observations of the activities of ball-rolling dung beetles, who push their dung balls backwards with their hind legs high on the ball, keeping their head down and employing it (when needed) as an extra lever on more difficult terrain. In other words, naming was derived from observation of the actual habits of the beetles themselves.

Not all collectors saw the benefits of Fabricius's peripatetic taxonomy. William Younge, writing to Sir James Edward Smith, who later founded the Linnaean Society, complained that Charles Louis L'Héritier de Brutelle – [a French amateur botanist who eventually rose to the highest ranks of the French Academy] … is stealing new plants and claiming them as his own, as Fabricius is doing in entomology. Fabricius has described 300 species just from [William] Jones' drawings without recourse to the insects themselves.' Younge warned Smith against 'foreigners, who seem to have their own ends in everything they do, and who deprive the English of the merit of discoveries', and complained that Fabricius had left London a week earlier 'loaded with new things, for we have a wonderful alacrity in giving to foreigners what we will not give to our own countrymen.'[7] Ironically, Smith himself was later accused of stealing names and descriptions of plants from other collectors, and eventually retired to Norwich to escape, as he wrote to Sir Joseph Banks, the 'envy and backbiting, or more nauseous cant, among authors and artists in the society of the great town.'[8] Being a gentleman collector at the time meant occasionally risking harsh societal censure.

We already know that the Egyptians observed dung beetles closely, but their imaginative vision was fixed on the relationship between gods and humans, and the mystery of human life and the afterlife. Everything in their world revolved around what lessons and secrets nature might unfold for humans. When dung beetles were observed within a Cartesian framework, however, the question of identification narrowed down to questions of what defined a species. Focus had shifted from humanity to the species. Linnaeus had chosen the form of the insects' wings as the key characteristic that identified an individual insect species, but Fabricius refined that to the insects' mouth parts. His reasoning was that every insect needed to eat, so similarities in mouthparts would be a useful indicator of relationships between species, bringing together groups, like dung beetles, that ate the same food.

There were two important basic thoughts in Fabricius's system. First, he distinguished between artificial and natural characters. He broke these down into those useful only in determining the species, and those showing relationships: *artificialis, quae classes et ordines, vel naturalis, quae genera, species et varietates docet* ('artificial with respect to classes and orders, natural with respect to genera, species and varieties').[9] An individual of a particular species will always recognise members of its own species when looking for a mate; but they have no concept of any other (higher order) categories into which we humans might group them. But while an insect does not need to know which family it belong to, humans do so to organise and retrieve information, and to discern the handiwork of evolution. Logically building these higher categories was part of Fabricius' enterprise. The system he established was not perfect, and he used terms that subsequently changed: classes for what we now call an order, and order for family; furthermore, he never actually named an order (i.e. our family). He understood that the genera (the plural of genus) should be grouped at higher levels within a logical system, but he was afraid that the time had not yet come for such sophisticated taxonomy. The genus, which he regarded as a natural combination of species, was most important to Fabricius and he offered some tantalising ideas about how variation between the species came into being:

> The enormous number and multiplicity of the species are due to an evolution of new species, partly as a result of hybridization among the species already existing, partly as a result of the facility of the external parts to attain new forms and thus to build up stable varieties which later evolve into species (1808).[10]

Fabricius also entertained concepts about the influence of the environment on a species, as well as sexual selection, believing that females preferred the strongest males, an idea that would emerge in the thoughts of Darwin in the not-so-distant future.

Although Charles Darwin was not the first person to consider why life on this planet was so diverse, he was the first to explain the mechanism that drove evolution. His grandfather Erasmus Darwin, in his book, *Zoonomia* (1794), suggested that all life had evolved from one common ancestor that had over time branched off into all the species we see today. He thought the transmutation of species was driven by competition and sexual selection, but he had no facts to support his theories. It was his grandson who would later provide the evidence.

Their thinking was not completely original. Lucretius (circa 99 BCE - 55 BCE) had written that 'many attempts [at life] were failures; many a kind could not survive; whatever we see today enjoying the breath of life must from the first have found protection in its cunning, its courage, or its quickness' – which sounds very much like natural selection at work. Indeed Anaximander (circa 610 BCE - 546 BCE) suggested in his text *On Nature* that life started out as slime in the oceans and eventually moved to drier places, but it was to take over 2 000 years before these ideas would re-emerge, together with the necessary evidence to support a solid theory of evolution.

As the natural world was being given a framework and order in the eighteenth century, social order was undergoing considerable upheaval. The disruption of the great chain of being in perceptions of nature had its parallel with a world where monarchs were no longer perceived as appointed by God. In France, being a scientist in such volatile times could be precarious, given that science remained within the purview of the well-educated and wealthy with few exceptions. Jean Baptiste Lamarck, the son of an impoverished aristocrat, was one of those exceptions. Unlike Antoine Lavoisier, a fellow aristocrat of considerable fortune who lost his head in the French Revolution, Lamarck was able to stay under the radar of the new Republic, and became the man who restructured the Jardin Royal des Plantes Medicinales into the Museum of Natural History in 1793.

Once the museum was properly established, French scientific research had a new home. There was broad overlap between areas of study, which meant that there was great fluidity between disciplines, with individual scientists studying an array of topics such as geology, zoology and botany. The overlap meant that theory was informed by a variety of emerging disciplines. Lamarck's ideas about evolution were based on his research work conducted on fossils excavated around Paris.

His colleague, Georges Cuvier (1769–1832) was also interested in such matters, based on his own research along the Normandy seashore. Cuvier was one of the scientists who undermined the concept of a great chain of being, positing instead that the animal kingdom was divided into four categories: Vertebrate, Articulate, Molluscs and Radiate. He developed a theory that the earth had gone through a number of natural and other cataclysms, leading to extinctions and the rebirth of groups of animals. Lamarck had an opposing view, and to support it he developed two laws of evolution. The first law was the 'use it or lose it' principle, which suggested that a used organ would grow stronger while a disused one would wither and disappear. The second law was that of 'acquisition', which stated that new generations of species would acquire modifications developed in their parents. Lamarck was ridiculed by Cuvier, even after his death, but Lamarck had the last laugh; in the contemporary field of epigenetics, his theories are finally being supported by exciting evidence that the environment can indeed influence the inheritance of features developed during the lifetime of an individual.

This was a brave new world that the natural philosophers (scientists) of the time were moving into, and one which was a product of a new approach to old questions about the world. The formalising of institutions and the naming and ordering of everything being gathered into museums was providing the framework for a revolution in perception. There were many unanswered questions – not the least of which was how did everything fit together?

In some senses, the naming of the world was still an enterprise in itself. Few were undertaking research in situ, and confusion as to the source of items identified bedevils researchers to this day. Fabricius, for example, would describe a species first by giving it a unique name, followed by a description of never more than two lines, and a reference to the locality of collection and the collector's name. This seems straightforward until one learns that often a named collector was merely the person who purchased the insect from a frequently unnamed commercial collector working in the wild. This meant that no one was really sure where something came from. The focus would thus inevitably be on the higher taxonomic grouping and species' name, rather than on environment.

It was in the nineteenth century (when environment was added to considerations of the origins of diversity, not simply as a theoretical concept following on the observation of a few creatures and places), that one of the greatest transformations in thinking about how the world was formed was allowed to emerge. The story of how this happened starts with the second voyage of the ship known as the *Beagle* to South America. This journey was undertaken largely as a result of political changes in the West. The defeat of the French under Napoléon at Waterloo in 1815 heralded the ascent of Britain as master of the seas, and the undisputed imperial power during the nineteenth century. Spain had lost its colonies in South America as a result of numerous factors (not least of which was its alliance with Napoléonic France) and newly independent South America represented territories of great interest to the expansion of trade for a self-assured Britain. The problem was that detailed geographic and maritime information regarding the continent was limited.

To support English trade with the newly independent former Spanish colonies, the English Royal Navy needed accurate knowledge about the coastline of South America. Acquiring that data was to be the main task of the *Beagle* on its second voyage. The circumstances leading to this ship's second voyage were unusual, to say the least.

On its first voyage to South America under Captain Robert Fitzroy (1805–1865), four unfortunate individuals from Tierra del Fuego were captured and taken back to Britain. They were given the rather random and strange names of Jemmy Button, York Minster, Fuegia Basket and Boat Memory (the latter dying from smallpox shortly after his arrival in England). Part of a questionable experiment as to whether it was possible to improve the lot of individuals perceived as primitive by teaching them Christianity and turning them into missionaries, the remaining three were taught English, outfitted and paraded around; however it soon became apparent to Fitzroy that no one was prepared to take any long-term responsibility for his captives.

Fitzroy realised that the Fuegians ought to be returned to their homes, and as he was the author of their removal, it was his responsibility to see to their passage home. His guilt essentially led to the second voyage of the *Beagle*. At first, the Admiralty were not interested in supporting his venture, so Fitzroy determined to raise enough money to finance the voyage himself. Eventually the Admiralty, with some encouragement from appropriately placed Fitzroy relatives, decided to sponsor the expedition and the project became official. When Fitzroy suggested that the presence of minerals in Tierra del Fuego was indicated by the wildly oscillating compass readings he had recorded on the first voyage of the *Beagle*, the Admiralty accepted his suggestion that a geologist should accompany the second voyage. The combination of taking responsibility for locating undiscovered resources and cold feet caused the Reverend Leonard Jenyns of Swaffham Bulbeck to withdraw from the job of naturalist, and this led to Charles Darwin being asked instead. It was one of those epic and fortuitous accidents that shift the compass of the world.

Charles Darwin boarded the *Beagle* in December 1831 as the official expedition naturalist, with a brief to spend five years travelling around the world observing, collecting and exploring geology and biology on a grand scale. He was not unsympathetic to the Fuegians whose return had led to the voyage, but while he did ruminate on their relationship to other humans, his real interest was the natural world. Dung beetles, although only a long footnote in his published 1845 journal of the voyage, feature in thought-provoking ways in his notes and collections. Darwin collected many beetles on his long journey, continuing a passion which caused him in later life to say 'Whenever I hear of the capture of rare beetles, I feel like an old war-horse at the sound of a trumpet'. Unfortunately, when he returned to England he could not find anyone willing to take his specimens. As a result of this, his collection was dispersed across a variety of institutions in Britain and Ireland, and so it is hard to get a representative vision of quite how extensive his collection really was. But his notes remain available, and these tell an intriguing story.

Darwin kept a list of the numbers of the different families of insects he found, and the places from where they were collected. Occasionally he made remarks on his collections, such as noting in 1832 at Bahia that the dung-dwelling histerid (a clown beetle) *Onthophilus* 'perceived the smell of human dung with singular quickness'[11], suggesting he was not above poking around in poo for interesting material, or using his own to lure in the wildlife. This may explain why the one beetle family that consistently drew the most comments was the *Scarabaeidae*. In northern Patagonia, he observed that the *Scarabaeidae* inhabited sandy hillocks near the sea and seemed to be living off the dung of what he called ostriches (they were actually rheas, large flightless relatives of the African ostrich). Darwin noted that he saw 'one [beetle] busily employed in pushing along a large piece with its frontal horns.'[12] It is notable that Darwin was not surprised by the sight of a dung beetle transporting its food the wrong way around, moving forwards instead of backwards like the majority of ball-rolling species. His description suggests he was watching a species of *Anomiopsoides*, which hold dry dung pellets in their front legs while running forward on their back two pairs of legs, an unusual way for a dung beetle to forage. As a native of Britain, Darwin would not have seen ball-rolling behaviour at home, as this only emerges in species occurring from the centre of

Europe southwards. Even as a famously keen observer and aficionado of the Coleoptera, Darwin's view of dung beetles was not broad enough to raise questions about the peculiarities of the scene he witnessed.

Nevertheless, Darwin was very interested in the presence or absence of dung beetles, particularly when associated with introduced livestock such as cattle and sheep. From his comments it is evident that he was thinking about resident beetle species adapting to new food opportunities, as he made notes on this theme more than once. He was also as interested in recording their absence as he was in actually finding dung beetles, making the following note in 1841 at Bahia Blanca: '*Coleoptera, Scarabaeidae*: no specimen found.' He went on to say:

> This absence of coprophagous beetles appears to me to be a very beautiful fact as showing a connection in the creation between two animals as widely apart as *Mammalia* and the *Insecta Coleoptera*, which when one of them is removed out of its original zone can scarcely be produced by a length of time and the most favourable circumstances.[13]

This comment is remarkable because it indicates that his assiduous collection of dung beetles was leading him to tentative conclusions about symbiotic relationships in nature and the co-evolution of species. It highlights why he was so interested in the presence of horse dung and beetles consuming it in Maldonado, alongside the Rio Plata near Buenos Aires. It was a world with an Alice-in-Wonderland 'curiouser and curiouser' quality; he noted that in Chile *Geotrupes*, while excessively abundant, had no large quadruped on which to depend for their food source. Even more inexplicably, on an island off the coast of Chile, he found large numbers of *Geotrupes* in gardens where 'dung (was) not directly present.'[14] Again in Chile, Darwin recorded *Phanaeus*, noting that 'on the opposite side of Chiloe, another species of *Phanaeus* is exceedingly abundant, and it buries the dung of the cattle in large earthen balls beneath the ground. There is reason to believe that the genus *Phanaeus*, before the introduction of cattle, acted as scavengers to man.'[15] He couldn't have guessed at the former presence of megaherbivores in the Americas, which had been exterminated by humans in the brief 14 000 years of our occupation of these continents.

A diverting image of Darwin wandering around in Chiloe (as he spelled it) emerges from his descriptions of his hunt for dung beetles, puzzled and fascinated that he was unable to find any despite diligently turning over every dung pat in sight. Dung beetles were mentioned throughout his *Beagle* journal, and adaptation to new food sources absorbed him. He was obviously thinking about how they might have adapted to new dung resources, as well as wondering how the beetles arrived in new locations. He was fascinated to see some living members of the *Scarabaeidae* bobbing along in the sea as the *Beagle* sailed near the mouth of the Rio Plata. He pondered over whether this might be a way that species found their way to new locations. It was a thought that occupied him in Tasmania and again in St Helena, where he observed dung beetles on remote islands; this led to a very long footnote as a conclusion to his observations about dung beetles on the voyage (see Appendix A).

The genius of Darwin's theory of evolution was a product not only of a remarkable mind, but also of his powers of observation, and the fact that he travelled across the globe and saw for himself the variety of responses of animals to their environments. It is arguable that he might not have reached the same conclusion if he had not travelled to so many destinations. He was able to make connections and accumulate evidence that had previously evaded other thinkers about evolution. This was a major triumph of empirical science, and one that changed the template for all subsequent thoughts about nature. Dung beetles rolling their balls and thriving or not thriving were undoubtedly part of his conclusions about adaptation, but their role in his thinking is generally downplayed compared to the observations he made about dung beetle horns in his book on sexual selection.

Darwin found the size and variation in dung beetle horns to be fascinating, but could find no evidence that they were used in combat, and so deduced that 'they have been acquired as ornaments.' This conclusion 'is that which best agrees with the fact of their having been so immensely, yet not fixedly developed, as shewn by their extreme variability in the same species, and by their extreme diversity in closely allied species. This view will at first appear extremely improbable ...'.[16] Darwin's view has been discounted, as it has since been observed that *Onthophagus* horns are used in combat between males where size does count.[17] Larger males have larger horns and win more fights, stationing themselves at the entrance to tunnels occupied by nesting females with whom they have mated. Their elaborate head ornamentation is used to block the tunnel to smaller rivals, who have to resort to sneaking through side tunnels to get to the females.

Darwin's observations on dung beetles and their role in his epic work reveals them as having a small but valuable part in his piecing together of answers to the puzzle of evolution. When he arrived home with his dried and pinned collection, however, his beetles were not in demand. He had difficulty getting his specimens identified, but wasn't prepared to leave them with the British Museum: 'I daresay the British Museum would receive them but I cannot feel, from all that I hear, any great respect even for the present state of that establishment.' The natural history department of the British Museum at that time was in a shambles and unable even to preserve their existing collections. The museum's 1833 annual report stated that of the 5 500 insects listed in the Sloane catalogue none remained (presumably consumed by museum beetles, the bane of all curators). Darwin's caution meant that his insect collection from the *Beagle* voyage ended up scattered among several museums, where he could find specialists prepared to work on his material.

The exercise of ordering and displaying specimens to the public was largely in the hands of individuals such as Richard Owen (1804–1892). He was a comparative anatomist and palaeontologist who became the first director of the Natural History Museum in South Kensington when it separated from the British Museum, a process that took over a century. Owen was responsible for the establishment of this new museum, which he saw as a place of research as well as an opportunity to educate the public. He also regarded the museum as a place to celebrate a large-scale microcosm of creation. Biological material was distanced from its sources and original environment, with the closest resemblance to the origins of the flora and fauna being recreated in dioramas or displays behind glass, with lions entangled by snakes, alongside a few plants and painted backdrops. Strange and antiquated when seen from contemporary viewpoints, it was nevertheless a world that fuelled a growing interest in nature among Victorians, whose passion for collecting drove many rarer species of flora and fauna to the brink of extinction.

The Reverend John George Wood (1827–1889), who became a tireless populariser of entomology, captured the shift in emphasis and perception in the preface to his book on foreign insects. He wrote that the object of the book was to show the 'great and important part played by insects in the economy of the world, and the extreme value to mankind of those insects which we are accustomed to call Destructives.'[18]

The intention behind the development of museums and research institutions in the metropolitan homes of colonial powers was one of popular education, and the ascendancy of Britain as an empire saw the expansion of its museums, with probably the most famous and ambitious project being that of Prince Albert, Queen Victoria's beloved consort. Having witnessed the impact on the British public of the Great Exhibition of 1851, he conceived the vision of a collection of institutions where useful learning in the physical sciences, engineering, manufacturing and the arts would be made available to the public. The outcome was the magnificent collection of museums and related institutions built around South Kensington in London. The project took 50 years to complete, but as an expression of the order, structure and values of the Victorian era little else continues to represent that period with such living clarity. Prior to that haphazard collections had been made (and sometimes lost), but they were managed with a level of eccentricity that (while amusing) did not serve science very well.

One could question what science meant at that stage, as 'the word *scientist* was not invented until 1833, and even then was slow to catch on. Both Michael Faraday and Charles Darwin refused to let themselves be labelled with the new term.'[19] In the same way that the naming of species helped to define and order the world, the movement from naturalists to scientists carried the same semantic dynamic. Science as we now know it was undergoing huge transformations, and the fields of the natural sciences seemed to waver between taxonomy (driven by the naming and ordering of the universe) and larger theories about the nature and the purpose of the world. This was not surprising, given that the physical evidence of the extraordinary diversity of the planet was still in the process of being found, assembled and catalogued. As this happened, those working with these discoveries began to make comparisons, draw conclusions and posit theories.

France was important in this regard. It had been the first country in Europe to simplify its institutions of higher education. The French Revolution of 1789 had put an end to previous structures, but they were reformed in 1808 under the single institution of the *Universite Imperiale*. What made this university system new and unique was the division of the arts and the sciences into separate faculties. This separation was a major change, subsequently adopted by the rest of Europe by the mid-nineteenth century.

It was part of the formalising of science as a discipline, which came at a period when France had pre-eminent researchers in the realm of the natural sciences. In the first half of the nineteenth century, however, science was served more by societies of like-minded individuals than universities; to be a member of many of the societies, one needed to be a man of some means (and, of course, women were in most cases barred from membership until the twentieth century).

Religion continued to play a significant role in the continuing belief that everything on earth was part of a divine God-ordained plan. Richard Owen, at the helm of the British Museum of Natural History, was committed to this view and as the director of a major museum he had the means to express his beliefs. When Darwin published his ground-breaking work *On the Origin of Species* in 1859, Owen became one of his most vociferous detractors.

One of the weak points in Darwin's argument, latched onto with alacrity by those who hated the suggestion that we might be descended from apes, was that he did not have an empirical way of proving how character traits were passed from one generation to another. He subscribed to the theory of the time: that parental characteristics were blended in the offspring. According to this, a tall and short couple would have a medium-sized child; red flowers crossed with white flowers would bear pink progeny. Variation would be lost and natural selection would have nothing act upon. Darwin knew it was a weak point in his argument, and it was one of the reasons he assembled such a large body of overall evidence in support of his theory which, while compelling, was flawed.

Owen initially received Darwin's work with interest, but the issue of humanity's relationship to the great apes was something that led to a complete breakdown in their relationship. Owen became so opposed to anyone who supported Darwin, that in 1871 it was revealed that he had been involved in a threat to end government funding of Joseph Hooker's botanical collection at Kew. Hooker and Darwin were firm friends, and shared ideas on speciation and evolution. In contrast to the worldly and powerful Owen, Darwin was a reticent individual who loved nothing better than working in the seclusion of his home, Down House, where (surrounded by his adored family) he could read and write and conduct endless and ingenious experiments.

Across the Channel in France, an equally reclusive man was also changing the story of entomology. No history on dung beetles and their journey into the world of science would be complete without the inclusion of the delightful Jean-Henri Fabre (1823–1915). He was traditional and tenacious on the subject of religion, which precluded him from ever accepting Darwin's theory, but like Jan Swammerdam before him he found a way to reconcile his religious beliefs with his observations. He wrote 'I do not believe in God, because I see him in all things and everywhere.'[20] For Fabre, the intricate perfection of what he so meticulously observed was proof enough of divine presence; humans could never produce anything as perfect, therefore the proof of God was in the perfection and working of nature.

Fabre was the son of poor uneducated parents who were neither successful peasants nor petit bourgeoisie. He had the good fortune to shine as a student, and due to the reformed educational structure in France, he escaped the family fate of remaining a landless peasant. He became a teacher and then, through further studies, a professor. His world was that of Provence in Southern France, latterly the hamlet of Serignan, and there, in what he called 'the incomparable museum of the fields', he conducted research into the life in the meadows around him. Fabre, like Darwin, closely observed creatures in their natural habitat. Where it was difficult, if not impossible, to observe the habits of an insect (particularly underground) he devised experiments in his study that allowed him to watch what insects did underground; this explains how and why he is given the accolade of being the first recorded person to observe the creation of a dung beetle brood ball. Unfortunately for him, Ulisse Aldrovandi appears to be the rightful claimant of this distinction, for having illustrated a dung beetle nest. Nevertheless, we have no record of Aldrovandi actually observing the creation of the brood ball. It might seem like splitting hairs, but it does allow Fabre to retain his spot on the dung beetle roll of honour. Fabre, apparently unaware of Aldrovandi's work, had long searched for information on dung beetle reproduction and had employed local shepherds and children to assist him in his search for evidence in the fields. It was in fact a young shepherd who found him his first pear-shaped brood ball. Unable to observe the buried brood ball in the field, Fabre created little observation boxes in his study, and it was via this voyeuristic device that he first observed the metamorphosis of a dung beetle, and recorded it for posterity.

His observations were the foundation for understanding the complex relationship between dung beetles and their balls of dung. We now know that a dung ball, fashioned by a beetle at a fresh dung pile, has three possible fates. For feeding, a medium-sized ball is made and quickly rolled away to be buried a few centimetres under the soil surface. This is consumed, alone, over four or five days, after which the well-fed beetle emerges and flies off to find a new dung pile. On the other hand, an amorous male might make a somewhat undersized ball, which he buries in a shallow, open tunnel, in which he does a head-stand at the opening. This posture allows him to release a pheromone to attract a receptive female, who, if she falls for his smelly charms, will then consume his nuptial dung gift. Finally, a brood ball is the most important role for a dung ball because it is the vehicle by which the beetle will propel its genes into the next generation. The brood ball in ball-rolling beetles is easily recognised as a perfect sphere, often the size of an orange. This impressive creation is rolled away by either a fertilised female, who can use sperm stored from a previous mating, or by a pair of courting beetles. Once buried underground, as deep as one metre in some species, the ball is reworked by the female, sometimes into two or three smaller balls. Into each of these she lays a huge egg, which may have half-filled her abdomen, and which will hatch into a larva. The larva completes its development inside its own ball, eating the inner wall and defecating onto it, only to re-eat the confection over and over again until reaching full size (about the size of a thumb joint) as it prepares to pupate. The magical metamorphosis of the body, in which the larva transforms from a worm-like eating machine into a glossy armoured beetle, is performed in a pupal chamber. This chamber is created from the final gut contents of the larva, which are ejected and plastered onto the inner wall of what remains of the dung ball. The larval faeces sets into a ceramic-hard substance that can be cracked only with a hammer. Safe inside this chamber, the corpse-like pupa lies motionless on its back while its whole body remodels itself for a new life above ground in the light of the glorious sun. In Africa, its escape is triggered by rainfall, seeping through the earth from soaking spring rains. This softens the faecal shell and stimulates the new adult to struggle up through the soil to the surface and a life of dung.

Fabre's observation of the brood ball was part of the further transformation of science into an evidence-based activity, and his influence on the field of entomology cannot be underestimated despite his very unscientific language. His vivid, dramatic and tender descriptions of insects capture the imagination – for instance, he describes the dung beetle pupa as an 'amber jewel'. He uses the writing of Horapollo, who recorded the Egyptian information about Khepri, as a point from which to analyse received information and wisdom, and describes the colouring of the nymph of the dung beetle in the most poetic terms, likening the emergence of the new beetle to a religious rite. He writes about its 'imposing raiment, blending the scarlet of the cardinal's cassock with the white of the celebrant's alb, a raiment that harmonises with the insect's hieratic character.'[21] He describes the drama of the now fully matured little beetle as a life and death struggle worthy of a Victorian melodrama: 'Will he or will he not escape from the natal cradle which has now become a hateful dungeon?'[22]

In Fabre's world, the intimacy and struggle for existence is rivetingly portrayed in language that would have most modern scientists in a froth of despair, but there is no denying the romance and delight he conveys on each page of his writing. His description of what he calls the banqueting chamber (the chamber being a hole in the ground and the food being a feast of dung) is an epic of imaginative 'scientific' writing: 'The table is sumptuously spread; the ceiling tempers the heat of the sun and allows only a moist and gentle warmth to penetrate; an undisturbed quiet, the darkness, the Crickets' concert overhead are all pleasant aids to digestion ... Who would dare disturb the bliss of such a banquet?'[23]

Yet although he observed so closely and uniquely the same struggle and adaptation to existence that captured Darwin's imagination, he did not agree with Darwin's interpretation. When he described the importance of rain for the emergence of the dung beetle, he did not make the leap into the world of adaptation and evolution so apparent to Darwin. He inclined rather to a Lamarckian worldview of evolution. His observations and writings were so compelling and thorough, however, that he could not help but win the respect of Darwin. Fabre likewise had great respect for Darwin, but drew the line at acceptance of his theory. He resorted to our friend the dung beetle to make his argument: 'Be we men or Dung-beetles, we are all created in the image of an unalterable prototype: the changing conditions of life alter us slightly on the surface but never in the framework of our being.'[24] He argued that there was no satisfactory explanation of heredity and kept the concept of a divine creator firmly in his worldview.

In the last years of his life, he was finally widely acknowledged as a pioneering entomologist and writer, but his world of science was a million miles from the formal and community-oriented scientists of today. Fabre was one of the last of an era of almost solitary scientists labouring away at their particular passion, starting to pry open the pages of the manual on how the world works. Delightful though his work still is, had he been in touch with other researchers (and if information had been shared more easily between the rather eccentric and individualistic scientists and naturalists of his era) he would have come across the work of three other scientists who in 1900 stumbled across Gregor Mendel's work on heredity.

It is not without irony that the resolution to the dispute over inheritance in evolutionary theory was found in the research results of a Catholic monk. Gregor Mendel (1822–1884) came from a poor Austrian family and the only place where a bright and bookish child could find an education was within the Church. He became a monk, and acquired the research skills and a place in which he was able to conduct his experiments on heredity. It was his punctilious work breeding successive generations of peas (conducted as an Augustinian friar in the quiet of the Abbey St Thomas in Brunn, Austria) that laid the foundation for an understanding of heredity and modern genetics. His work was one of the essential missing pieces of evidence needed for Darwin's theory of evolution, as it explained how characteristics of parents were passed on unaltered to successive generations. Mendel showed that character traits (genes in modern parlance) are not blended in offspring; they remain independent and can be separated in later generations – pink flowers from white and red parents can be cross-bred to yield white and red offspring. Unfortunately, neither he nor Darwin lived to earn the recognition or vindication the completion of this part of the puzzle would offer to the evolutionary enigma. From the point of view of scientists who wanted to believe in the agency of an all-powerful creator deity, the extra years of doubt and lack of evidence offered the opportunity to indulge in sustained shredding of Darwin's theory.

Mendel had worked more or less at the same time as Darwin, but he was in Austria, and when the results of his eight years of meticulous research were published in the obscure journal of the Natural Science Society of Brno in 1866 in German, they made hardly a ripple. Ironically, Darwin owned a book, *The Plant Hybrids* by Wilhelm Olbers Focke (in German), that mentioned Mendel's pea experiments. Although Darwin wrote his name on the title page, he did not read the book as evidenced by the uncut pages.[25] Books were bound with the outer edges of the pages connected, to be later cut by the binder, or reader, and Darwin had never turned the pages of this one. Not surprisingly, Mendel's work remained largely unnoticed in the salons of Victorian science, and so the missing information as to how heredity works lay dormant for 34 years until 1900 when three botanists (Hugo de Vries, Carl Correns and Erich von Tschermak-Seysenegg, each working independently) rediscovered Mendel's work.

By this stage cells and chromosomes were sufficiently understood for Mendel's work to have a physical explanation. Fabre, the tenacious explorer of the microcosmos, was still working in 1900 (indeed he lived to 1915) but his little world of dung beetles in underground banquet halls in Provence remained unsullied by the new energy heralded by the twentieth century. He belonged to another time, in which information and ideas circulated slowly. The era of the solitary scientist was almost over, but not without enormous payoffs. Fabre not only brought the intense life of insects to a wide audience, but also engendered generations of youthful entomologists, while Darwin changed the entire future of science. The dung beetle played a significant role in shaping the thinking of these two remarkable scientists.

CHAPTER FOUR

Colonising insects

THE NATURAL SCIENCES EMERGING in Europe and Britain during the nineteenth century took on a different trajectory to that in the colonies. Settlement and expansion of growing populations had brought massive disruptions and changes to 'new' environments. Whereas the old world had centuries of established farming practice and land utilisation to support human needs, the territories of the new world were experiencing a tsunami of change. With this came new problems and new pests, as well as freshly qualified individuals equipped to respond to these problems. They called themselves scientists, as opposed to natural philosophers (a term Darwin and many of his generation preferred). In response to a taunt by the poet Samuel Taylor Coleridge (at a meeting of The British Association for the Advancement of Science in 1833) that the term 'natural philosopher' needed to go, because philosophers advanced humanity through thoughts rather than actions, William Whewell (1794–1866) coined the phrase 'scientist' as an analogy with artist .[1] Whewell was a polymath Cambridge scholar who excelled in both mathematics and poetry. He was also a wordsmith, generating the terms 'physicist', 'linguistics' and 'astigmatism', among others. Spawned from the humanities, the word and work of scientists has come to dominate our lives, whatever we choose to call its practitioners.

Following the American Civil War in the 1860s, agriculture expanded in North America on a huge scale. In 1869, there were just over two million farms in the United States. Forty years later, there were over 575 million farms, and land under cultivation had more than doubled to over 840 million acres. Most of this development was in the Midwest and West where millions of acres of tallgrass prairie land were brought under the plough – often with disastrous ecological consequences.

As agriculture grew in the post-reconstruction economy, productivity at first expanded. The virgin land was so productive that North America's farmers became victims of their own success as produce prices dropped. But land cannot be farmed without replenishing the soil, and poor soil produces poor crops, which in turn are more susceptible to attack from disease and insects. Monocultures (particularly cotton) expanded, and the insects preying on crops also increased in variety and number. The boll weevil, cotton's nemesis, was poised just across the Mexican border to wreak havoc on the cotton kingdom of the American South.

North American entomologists understood that most imported crop pests would have natural enemies in the territories from which they originated: as early as 1855 Asa Fitch (1809–1879), state entomologist of New York, advocated the importation of European parasites of the wheat midge *Sitodiplosis mosellana* (which causes damage by larval feeding on the developing kernels), but nothing came of his attempts. Although this new cohort of scientists were as passionate and knowledgeable as their modern counterparts, the lot of entomologists encouraging biological control was summed up in 1866 by Benjamin D. Walsh. Addressing the contemporary infatuation with new-fangled machines and progress, he wrote that if someone professed to have discovered 'some new patent powder pimperlimpimp, a simple pinch of which being thrown into each corner of a field will kill every bug ... people will listen to him with attention and respect. But tell them of any simple common-sense plan, based upon correct scientific principles, to check and keep within reasonable bounds the insect foes of the Farmer and they will laugh and scorn you.'[2]

Despite this attitude, which still prevails in the form of modern addiction to insecticides, nineteenth-century American state entomologists resorted to biological control of pests with many spectacular successes. One of the founding fathers of economic entomology Charles V. Riley (1843-1895), who became the second federal entomologist in the US, preferred the biological control route. He and his colleagues actively searched for natural predators and enemies of insect pests.

Biological control can take different forms. Usually a natural enemy (which can be a fungus, bacteria or other foe of a problem insect or plant) is identified in the country or territory of origin of the pest, checked for safety (to ensure that it does not attack non-target organisms), then bred in its new home in sufficient numbers and released to kill off the invasive pest. Predators can also be used, and they are often successful in controlling a target insect. In the case of dung beetles, their destruction of the breeding ground of filth, flies and other pests is slightly different. The beetles generally do not kill the pest itself, but their rapid monopoly and consumption of dung limits the breeding capacity of these pest species. Biological control can utilise a multi-pronged approach using more than one enemy, the goal being to use natural enemies instead of chemicals to suppress a problem insect or disease that has undergone a population explosion, and to restore some sort of natural balance within a particular ecosystem.

As an early devotee of biological control, Riley was part of an unusual array of individuals: keen entomologists who differed from the scholarly naturalists engrossed in cataloguing the world from the dim recesses of museums of Europe and Britain. By contrast, the colonial breed of entomologist seemed to be a mix of adventurer, naturalist, detective and practical problem-solver. Riley, who was to have considerable influence in creating the field of applied entomology, was typical of this small group. Born in London, he went to school in France and Germany, where he pursued his love of insects and drawing and planned to become an artist. At the age of 17 he went to the US and found work on a farm in Illinois about 50 kilometres from Chicago. It was there that he first became aware of the damage caused to crops by insects, and wrote a piece about it in *The Prairie Farmer*, a local agricultural magazine. At 21, he moved to Chicago to write full-time for the same magazine, becoming the reporter, writer, artist and editor of the grandly titled entomological department at the magazine. Riley's writing attracted the attention of the Illinois state entomologist Benjamin Dann Walsh, which led to him being appointed the first entomologist of the state of Missouri in 1868.

For the next nine years, in collaboration with Thaddeus William Harris, B. D. Walsh and Asa Fitch, Riley published nine annual reports considered to be the foundation of modern economic entomology. Riley's work on the pollination of the yucca plant by the yucca moth was of great interest to Darwin because it provided evidence of co-evolutionary relationships between plants and insects, resulting in their survival being dependent on their mutual survival – what is grandly known as obligate mutualism. (In this case, the moth's larvae use only the seeds of the yucca plant for food, and the yucca plant is pollinated only by the yucca moth). Riley's focus, however, was on trying to find solutions to the runaway insect explosions in the newly developing American farm lands. His love and understanding of insects meant that he advocated biological control above other quick-fix methods, which influenced those who worked with him. Among them was Albert Koebele (1853–1924), the man who was to bring dung beetles to the island of Hawai'i, and of whom we shall soon hear more.

Hawai'i has the distinction of being the first place where dung beetles were deliberately introduced as a means of biological control. Islands in general are particularly susceptible to the ravages of foreign imports and Hawai'i, despite being one of the most geographically isolated places on earth, has been no exception. The lush flora and fauna observed by Captain Cook in 1778 when HMS *Resolution* set anchor at those far-flung islands was by no means representative of a pristine world. The 'canoe plants' of the original Polynesian settlers, which included everything from taro to bamboo, were well established by the time the first Europeans arrived. Hawai'i and its neighbouring islands had been originally settled by Polynesians between 400 and 500, and had remained in comparative isolation until Captain Cook's arrival. Cook brought (in addition to other items) a male and female boar, probably sourced from Indonesia where he had picked up supplies before heading out into the vast Pacific Ocean. They were the first of what was to become a flood of food-oriented introductions.

A list of crops introduced to Hawai'i gives a sense of the laboratory-of-change experiments that islands such as Hawai'i became. By 1813, coffee and pineapple were reported in the gardens of Don Francisco de Paula Marin, a Spanish advisor to King Kamehameha I, and in 1824 the mango disembarked, courtesy of Captain John Meek. These arrivals were followed by a steady stream of potentially valuable economic crops. The introduced crops and livestock led to the first private attempts to control pests by biological means. In 1865 Dr William Hillebrand, a physician, naturalist and author of the *Flora of the Hawaiian Islands* (1888), introduced the mynah bird (*Acridotheres tristis*) to feed on armyworms infesting newly cleared pastures. Armyworms are actually the caterpillar of a moth, but the intimidating common name of these voraciously destructive creatures perhaps explains why such a wildly inappropriate (and noisy) solution was embarked upon.

Hillebrand's ill-advised introduction was but one of a cascade of other environmental disasters, not least of which caused the decline of native birds and the spread of the pretty but invasive weedy shrub, *Lantana camara*. Both mynahs and lantana now feature on lists of the ten most invasive species on the planet, and Hillebrand's well-intentioned intervention is probably a classic example of how and why biological control had to evolve a complex set of protocols and limits to prevent such repeats. Fortunately for Hawai'i, measures against these two species were taken at an early stage and (although still a problem) they were mostly brought under control. Another reckless introduction was the Indian mongoose, *Herpestes edwardsii*, which was imported from Jamaica by sugarcane plantation owners in 1883 and 1885. The mongooses were supposed to provide rat control; the rodents themselves were an introduction from South East Asia, courtesy of the early Polynesian settlers. But mongooses are active during the day while rats are mainly nocturnal; so the mongooses turned instead to more vulnerable targets (ground-nesting birds and other local delicacies such as turtle eggs) to the point where most of their favoured prey became endangered, and almost 30 bird species extinct. As a laboratory of humanity's penchant for speeding up evolutionary pathways in terms of annihilation, Hawai'i certainly qualifies as a notable example.

It was not only the flora and fauna that were transformed; the human population was also distorted by the newcomers. As in the case of so many isolated islands, the arrival of Western diseases had a terrible impact on indigenous inhabitants who had no resistance to them. Only seven years after Cook's visit, sailors from a passing French frigate found that most of the population had contracted tuberculosis and venereal diseases. The new settlers, particularly the Americans and the British (as well as the Japanese) founded some of the major food crop plantations. They brought their own religions, diseases and variations of greed, rapidly claiming much of the land (primarily the wetter East, onto which they imported labour required to work the plantations). New crops and imported livestock flowed in along with the people. The livestock arrived with their own hitchhikers, one of which was the particularly nasty bloodsucking horn fly – first reported in 1898, although it may well have arrived as early as 1793, when the first cows were imported.

The horn fly, a small Mediterranean muscid fly, which Linnaeus identified and most appropriately named, *Haematobia irritans*, is about half the size of a house fly. It derives its name from the fact that it often clusters in dark rings around cows' horns. The flies more commonly congregate between the shoulder blades, as well as on the flanks and belly, where the flick of a tail has absolutely no effect. These tiny travellers rapidly became a significant livestock problem. They spread to all the Hawaiian Islands, and can now be found from sea-level to two thousand metres.

The cattle definitively blamed for having brought this scourge originated in California; in two trips between 1793 and 1794, Captain George Vancouver brought a total of eight cows and four bulls to the main island. Two beasts expired almost on arrival, but the others survived and flourished, and their numbers grew exponentially under a taboo placed on the slaughter of cattle by King Kamehameha I. The king's successors had cause to regret their ancestor's decree, as by 1830 the offspring of these few pioneer cattle had become feral and so abundant that Kamehameha III ordered a massive slaughter of the feral herds.

Obviously lacking natural enemies, the horn flies prospered from the increase in cattle, and by 1898 the hitchhiking pests had become a significant problem on the islands. The horn fly caused economic losses to both the meat and dairy industries, and there seemed to be no local solution to a concomitant problem – an ever-growing amount of dung. There was no creature on the islands to process these faeces, and so they piled up, creating a breeding haven for an exploding horn fly population. The first attempt at biological control of the horn fly was the introduction and release of a predatory histerid beetle, *Hister* sp. from Puerto Rico in 1898; it is a relative of the clown beetle we encountered through Darwin. This was followed by the introduction by 1910 of 12 more species of natural enemies, mostly parasitic wasps (*Hymenoptera*). A big, black shiny beetle, *Hister* looks the part with its impressive jaws and irascible nature, but it failed to deliver. The pimperlimpimp one-stop plan was obviously both alluring and persistent.[3]

By the early 1900s, the horn fly problem had grown to the extent that Albert Koebele (by now the main entomologist on Hawai'i) was focused on finding an alternative solution to the problem. Koebele's rise as an entomologist had been rapid. Born in Germany, he had a love of natural history but little education. He left Germany in 1880 in response to a failed romance and set off for a new life in the US. A year later he was appointed to the dizzying height of assistant entomologist for the United States Department of Agriculture, based in Washington DC. However, he disliked life on the East Coast and managed to be transferred to Alameda, California in 1885 where the Australian fluted scale insect was decimating the fledging citrus industry.[4] In 1888, Koebele sailed to Australia where he collected a furry little ladybeetle, the Vedalia beetle.[5] Once productively established in California, the beetle was so successful that by the summer of 1889, the grateful citrus growers presented Koebele with a gold watch for his efforts (along with a pair of diamond earrings for his wife, suggesting that he had found more than just beetles in Australia). His venture is widely regarded as marking the beginning of modern biological control, where a host-specific agent is selected that will only attack its target pest and nothing else. Ironically, it may have been a motivation for the 1916 introduction of the Asian ladybeetle to control aphids (plant lice) into the US. In contrast to its predecessor, this large, voracious and indiscriminate predator eats other ladybirds and occasionally indulges in cannibalism. Releases of this insect continued until the mid-1970s – the pimperlimpimp dies hard. The Asian ladybeetle is now regarded as an invasive pest in much of the world.

After resigning from the US Department of Agriculture in 1893, Koebele was then appointed entomologist of the provisional government of the Republic of Hawai'i, and given responsibility for biologically controlling the many species of immigrant insect pests flourishing in their new home. Koebele settled on the importation of dung beetles into Hawai'i as a solution to the horn fly problem. His choice reveals not only the very experimental nature of early economic entomology, but also the chasm between what was happening in the New and Old Worlds. Fabre's work on scarabs was published only in 1918, and was the first known account of the full life cycle of a dung beetle. Despite a lack of detailed knowledge about dung beetle biology, they were nevertheless selected to resolve the horn fly problem. Koebele's lack of knowledge about the specific beetles also suggests why he chose to introduce a number of species. This was the kind of cavalier experimentation that prevailed in an era when there were so many gaps in our understanding of the natural world that optimistic guesswork was bound to be part of any solution. What made the importation of the beetles different from other biological control projects was that the primary target was not the pest itself, but its breeding ground: the dung. Less was likely to go wrong should the dung beetles run amok, as they were tied to this largely unpopular food source.

In 1909 the first dung beetles were imported from Europe to Hawai'i, along with quantities of material from which dung beetles were reared; these in turn may have brought a host of much less visible new living organisms to the islands. A year later, Koebele travelled to the US and made large collections of beetles in California and Arizona from which he introduced at least 'six or seven species of dung beetles … into the Hawaiian Islands.'[6] Two species of parasitoid[7] wasps reared from material sent by Koebele proved to be more useful in destroying the horn fly.[8] In total, 29 species of scarab dung beetles were eventually introduced and released. A few species that came under their own steam with the flies were apparently successful in establishing a toehold in the islands. Less than half of the deliberately introduced species persisted, with the survivors being mostly of African origin, and having been introduced at later dates (See Appendix B).

What is most striking about the list of introduced dung beetles are the dates of their importation – which reveal that none of Koebele's early introductions survived. The most successful introductions were initiated in the late 1950s by Cliff Davis, Chief Entomologist of the Hawai'i Department of Agriculture, assisted by Noel Krauss, an exploratory entomologist. Krauss paid attention to the symbiotic relationships between species and focused on natural enemies of the various insect pests and weeds of Hawai'i. He travelled widely and collected dung beetles such as *Onthophagus gazella* and *Liatongus militaris* in Africa, and *Oniticellus cinctus* and *Onthophagus sagittarius* in Sri Lanka. *Onthophagus gazella* from Hawai'i, translocated a decade later into Australia, would go on to open the doors for the massive Australian Dung Beetle Programme, winning favour with farmers who were impressed by its industriousness. *Euoniticellus intermedius* is another one-centimetre-long little workhorse from Africa that has now colonised the entire continent of Australia, excepting the southern coastline. The rolling dung beetle genera had mixed success, with only the two American *Canthon* 'tumblebugs' listed as surviving and the African *Neosisyphus* failing to establish itself in Hawai'i. *Canthon humectis*, introduced into Australia in 1969, also disappeared. The most surprising Hawaiian failures were the four species of *Onthophagus* brought in from Australia around 1921. Given that there are no kangaroos in Hawai'i, and the later flood of dung beetle species that poured into Australia in the 1980s, those efforts now appear naively optimistic. Maybe the Hawaiians were hoping that a little of the Vedalia beetle antipodean magic might rub off on their beetles.

The relative lack of success of Koebele's early dung beetle introductions point to some of the complexities of biological control. Frequently when biological control agents are introduced, they fail to reproduce and survive in their new habitat because some unknown element is missing. It could be a mutualist (like the yucca moth's host), something in the climate, or even a mate – some introduced species disperse before mating, resulting in such a low density of individuals in their new home that they simply can't find one another.[9]

Nevertheless, there are many examples of introduced (deliberate and otherwise) insects that have thrived on a new continent and, given that dung beetles eat dung, it no doubt seemed logical that they would fall into their new Hawaiian home and onto its abundant food supply with alacrity. Nevertheless, even though virtually all dung beetles enjoy dung, they do not all thrive at the same temperatures, environmental conditions, nor even particular varieties of dung. To complicate matters, not every dung beetle lives exclusively off dung; some flourish eating rotting fruit or vegetation or even dead animals. They are, within their subfamily, confusingly both specialists and generalists, keenly sensitive to the environment in which they survive.

For instance, in South America, particularly in the Amazonian forest, the creation of grasslands for cattle grazing has resulted in a significant loss in dung beetle populations and diversity. Where howler monkeys operate on the edges of forests, next to cleared land that has then been subsequently abandoned, dung beetles have been found to be critical to the process of natural reforestation. The howler monkeys have what amounts to communal lavatories: the end of a favourite branch, from which they drop their faeces in one large steaming pile. The local dung beetles process and redistribute the seed-rich droppings, which ultimately results in regeneration of the forest as it slowly creeps back into the disturbed land. However, this does not mean that those same dung beetles will range further than the forest edge and food supply they know. Where trees have been cleared, there has been a concomitant reduction in dung beetle numbers. The same effect has been seen in Australia, where indigenous dung beetle numbers decline on land that has been cleared because dung beetles simply do not walk or fly out of forested areas and adapt to a new supply of dung. That would require an instantaneous reversal of millions of years of evolutionary adaptation.

Dung beetles are sensitive to far more than just the flavour of their food, and in places such as the South American rainforest they have revealed delightfully high levels of specialisation. Dung beetles have been found perching high in the tree canopy, opportunistically waiting to harvest gluey gifts evacuated from tree-dwelling creatures which stick onto leaves as they fall. The beetles fly to this dung then roll it off the leaf surface, falling with it to the forest floor. This ensures that they get their first bite of the pie in a very competitive environment. Other species simply hang out in the genital and anal fur of arboreal monkeys (where some researchers said they 'glisten like jewels'[10] – maybe the irritation is worth the attention). The beetles fall with the dung as the monkeys defecate, claiming their prize as it hits the dirt. Other more patient species calmly sojourn in the fur of sloths, where they await the slow but deliberate trip to the ground to defecate. Their patience pays off in being first served at this weekly event. There is even a tiny dung beetle (*Zonocopris*) that feeds exclusively on snail slime, living apparently permanently on the mantle of live snails.

These complexities of dung beetle life are explored further in later chapters, but what is pertinent at this point is that the early importations of dung beetles into Hawai'i revealed that having a supply of dung is not always enough to satisfy a dung beetle. The gradual understanding that far more needs to be known for a successful transplant dawned at a later stage. Matching beetles to climate, soil type, rainfall patterns and so on became more obvious as their native ranges were described from collection records, while the failed introductions provided valuable pointers to what might be constraining a species in its native habitat. Dung beetles in Hawai'i in the 1950s were a case study in action.

The research that would yield explanations about the complex variety and adaptations of the beetles was still in the future, and the information necessary to solve the pressing problems of an island ecology under stress from new and successful invaders did not exist. Ecology was not even recognised as a concept when the first batch of dung beetles was introduced into Hawai'i. If anything, there was a mechanical approach to nature which fitted well with the enterprise of ordering and structuring it. There were no canaries in the coal mines sounding off warnings about major dangers to the natural world. Science was more preoccupied with the fundamental nature of matter, and working towards an understanding of how everything worked at an atomic and chemical level. Ernest Rutherford's alleged comment that all science is either physics or just stamp collecting captures a certain zeitgeist; however, it also reveals the same blindness that had prevailed in the early days of scientific nature paintings when isolated plants and insects were recorded separately from their environments. There was a degree of human hubris in the excitement at the discovery of the apparent core of the inner workings of the world, and it further fuelled the mechanistic approach to humanity's interaction with the natural world.

Economic entomology, for example, reflected a view of insects as primarily problematic. They were not viewed in the context of an overall system or as signs of an environment out of kilter, but rather as specific threats to human food crops. The individuals addressing the problems were technically fighting fires: even if they understood the causes of the problems, they knew they could not turn back the clock: the insects were present and solutions to their exploding populations needed to be found. Entomology itself was a relatively disparate and disorganised field. Once the province of predominantly amateur naturalists, by the turn of the twentieth century there was a shift away from entomology as an individual's private indulgence. Nevertheless, it was only in 1910 that the first international entomological conference was held. At this conference, economic entomology was tellingly represented in the section of economic and pathological entomology. Insect infestations were considered problematic and a pathology, rather than symptoms of a world of massive ecological disruption.

Biological control, a slow and frequently hit-and-miss affair, did not have the same appeal as or success rate of poisons. Farmers needed instant solutions as demand increased and output grew. It is not surprising to learn that in 1929 American fields and orchards had almost 30 million pounds of lead arsenate and calcium arsenate spread over them. At the time, calcium arsenate was the number one pesticide used by growers to fight the cotton boll weevil. These were the same fields from which our first American record of a tumble-turd came.

Agriculture needed solutions to the pests and diseases that preyed upon its crops and animals, so a considerable arsenal was developed to address the problems. Sodium chlorate, sulphuric acid, or organic chemicals derived from natural sources were some of the main constituents used in pest control formulations up until the 1940s. Ammonium sulphate and sodium arsenate were used as herbicides, while others such as nitrophenols, chlorophenols, creosote, naphthalene and petroleum oils (used for fungal and insect pests) were by-products of coal gas production and other industrial processes. The drawbacks of many of these products were their high rates of application and their lack of selectivity (they poisoned everything), as well as their phytotoxicity (they damaged the plants they were supposed to protect). They also proved to be not particularly effective, and became less so as many insects built up resistance and their numbers resurged.

Darwin's theory of evolution re-emerged only in the 1930s, but began to gain serious traction as 'the Modern Synthesis', which combined his ideas with contemporary genetics. But this momentous scientific shift was overshadowed by the effects of the 1929 stock market crash and the huge social disruption that followed in its wake. The first major environmental disaster in the US, the dustbowl that rendered vast tracts of the mid-West barren, was seen as a singular event rather than the first warning of how ill-advised contemporary farming practices had become. The dustbowl swept through a vast swathe of land encompassing Oklahoma, Texas, some of Kansas, Colorado and New Mexico, and covering approximately 150 000 square miles. It corresponded roughly with the grasslands that had been ploughed up for new farms after the Civil War of the 1860s. Drought struck the area from 1934 to 1937, and in the absence of any grass to anchor the soil, enormous swirling dust clouds, driven by the winds that annually sweep across central North America, choked most agricultural life. Approximately two and a half million people, 60 per cent of the region's population, left the area, mostly heading westwards to settle elsewhere. One silver lining was the formation of the Soil Conservation Service in the US in 1935. But even though the state assisted in restoring much of the dustbowl land to relative viability, farmers repeated the same mistake of ploughing up grassland for wheat during the Second World War when grain prices soared.

The dustbowl was one of the first visible signs of the outcome of an interconnected world transforming on a rapid and vast scale. Meanwhile, the tentacles of empire elsewhere were resulting in similar (although far less well-documented) transformations, such as the destruction of much of Malaysia's forests for the creation of rubber plantations (a process that has accelerated in the past decade with the establishment of palm oil plantations).

The Second World War saw even further environmental destruction, as well as the beginning of the nuclear age and the end of the British Empire. One outcome of the war was a world filled with even more chemicals. Dichloro-diphenyl-trichloroethane (DDT) epitomises those formidable compounds. First synthesised by a German chemist in 1874, it was reintroduced by the Swiss pharmaceutical company Geigy after their researcher Paul Müller (who had been searching for an insecticide to kill clothes moths) noted the rapidity and effectiveness with which it killed insects. The introduction of DDT as a weapon in the war against insect-borne diseases could not have been better timed. Allied troops were succumbing to malaria and typhus (carried by mosquitoes and lice respectively) and other insect-borne diseases. The Allies loaded DDT into mortars as they advanced through Italy, taking the soft underbelly of Europe. By blowing the powder into their forward positions, they eliminated the malarial mosquitoes breeding in flooded shell holes. Allied soldiers were issued with DDT-impregnated underwear capable of killing a typhus louse after twenty washings. Teams of Allies and local workers managed to fumigate the population of Naples over the winter of 1943 by using powder-guns to blow DDT the dust up their sleeves, down their trousers and into their hats. More than two and half million people were exposed to the insecticide: it probably saved their lives, and did them absolutely no harm – the mammalian toxicity of DDT is lower than table salt. However, it persists in the environment to the extent that we are still finding it in our food chains more than 40 years after it was banned; meanwhile evolution, in all its artfulness, rapidly sidestepped DDT by naturally selecting for insects resistant to it. DDT-resistant houseflies were discovered two years after the war. The imperial dream was coming to an end, but its transformation of the planet was one of enduring and continuing environmental change.

A more positive outcome of the post-war economy was the growth in education. New universities were founded, and the numbers of students grew rapidly. The great post-war expansion of universities took place in the late 1950s, while the numbers of returning servicemen and women registering at universities continued to increase. A by-product of the changing political landscape of the post-war years was migration. Young, educated individuals, wanting to build a life away from the geopolitical landscape of a reconstructing Europe divided between Communist and capitalist control, immigrated to different continents.

One of these individuals was George Bornemissza (1924–2014), a Hungarian scientist who studied science at the University of Budapest and obtained a PhD in Zoology at the University of Innsbruck in 1950. At the end of that year, he immigrated to Australia. There he saw the impact of generations of settlers and their cattle on the land, and could not fail to observe the famous Australian habit of people constantly brushing pesky flies from their faces.

The flies were a long-term unwanted product of the arrival of the First Fleet in 1788, when Captain Arthur Phillip offloaded a bull, four cows and an 'Indian Zebu' bull calf (bought in Cape Town, South Africa and all transported on his ship HMS *Sirius*). They were of mixed breed, with Brahmin humps, short tails and long horns. They had not been intended as food, but were rather brought to transport goods and to reproduce.

This original small herd was left in the care of convicts, who later allowed the bulls and three of the cows to wander out of their enclosure. They were assumed to have been killed and lost forever, but in 1795 convicts saw 61 cows grazing about 60 kilometres from present-day

Sydney. The hardy South African cows had somehow survived and multiplied. Captain Phillip, now Governor, named the area Cow Pastures and kept it as a strictly controlled area for government-owned cattle. Further cattle sourced in South Africa had arrived in 1791 and again in 1792, but the sea journeys between stops were long, and the numbers that made it to Australia were small. It was in 1795, the same year that the initial escapees had again been sighted, that a further one 131 head of cattle arrived from India. In subsequent years, another 296 cows from South Africa were imported to become the foundation of the government herd. These were the earliest arrivals on Australia's shores of what was to become one of the nation's richest agricultural enterprises, as well as one of its many ecological problems.

Australia was an extraordinary cornucopia of mammals and flora that had evolved in long isolation. It would take Alfred Wallace (Darwin's specimen supplier, co-author and correspondent) to explain how Australia and its flora and fauna had evolved. Without the benefit of a theory of continental drift, Wallace nevertheless observed and documented an imaginary biological line in the East, between Oriental and Australasian ecosystems. This seemed to separate abruptly one set of flora and fauna from the other. Based on his observations, made during extensive travels, Wallace came to a similar conclusion to what Darwin had published in *On the Origin of Species.*[11] What we now call biogeography was simply a record of inheritance. As organisms colonised new habitats and their old ranges were divided by mountain ranges or other barriers, they took on the distributions they largely still have today and over time adapted and developed into different species to suit those habitats. Australia is on the other side of what is now known as the Wallace Line, a biogeographical region different from the five other regions on the planet, which explains its unique flora and fauna.

Australia's people and its many unique creatures had evolved in glorious isolation (for at least 60 000 years, in the case of the people) but few, if any, of the indigenous plants and marsupials were regarded as useful or even particularly desirable by the early settlers (who imported their preferred food types instead). The resulting sheep, cattle, pigs and rabbits each brought their own problems.

Sheep in Australian paddocks were, as Don Watson (an Australian naturalist historian) described it, never more than a drench away from resembling a field hospital. This was because of the dipping used to prevent fly strikes (which were caused by blowfly maggots eating into the flesh of the sheep's dung-contaminated tails). This entomological problem has left New South Wales with 40 000 sheep-dip sites contaminated by arsenic. The rabbits that had been brought as a food source with the First Fleet had long since escaped, multiplying into a virtual biblical plague. This plague only began to be reduced with the introduction in the 1950s of the myxoma virus (which was used as a very successful biological control) as well as the use of the fumigant chloropicrin, which had first been used in the First World War on humans. All of these difficulties were a by-product of the introduction of food sources for humans, and the herds of cattle created a set of environmental problems of a different order.

The gold rushes in Australia in the 1850s had increased the overall population of immigrants by well over half a million people. They all needed to be fed, and everything from beef to potato production expanded in response to this new market. Sidney Kidman (later Sir) opened a butcher shop to process meat, and at the same time acquired huge tracts of land across the length and breadth of Australia on which to breed cattle. Transport for the cattle was in short supply, so they had to walk long distances to pasture or the abattoir, resulting in considerable losses. This led to a greater focus on breeding in order to produce cattle that could endure the harsh Australian climate and exertion of long treks to market. Australia had started to export cattle as early as 1829 and this business expanded further in the 1880s when the Northern territories were opened up to cattle farmers responding to the growth in demand for meat from the Asian markets. It was apparently also from the Northern territories that the greatest scourge of cattle came. Although it is hard to record exactly when a particular fly arrives anywhere on earth (except maybe Hawai'i), it appears that the buffalo fly might have been introduced into Australia from Timor in the 1890s. With ever increasing livestock numbers, the indigenous bush fly (*Musca vetustissima*) and the buffalo fly found fertile breeding ground in the copious amounts of cattle dung. Even though there are over 350 species of indigenous dung beetles in Australia, they are adapted to marsupial dung, which is dry and fibrous compared to the fine, wet dung of ruminant livestock. This meant that local dung beetles largely ignored the abundant new dung bonanza.

On Darwin's visit to Van Diemen's Land (Tasmania), he had found four species of *Onthophagus*, two of *Aphodius*, and one of a third genus, all which he commented were very abundant under the dung of cows. He was fascinated by this discovery, which suggested to him an example of adaptation in a relatively short time. This rapid adaptation on the island did not take place on the Australian mainland, however, where three of the same four species he had seen still associated themselves exclusively with the dung of indigenous animals.[12] Although Darwin had noted something extraordinarily interesting about how rapidly evolutionary adaptation could take place within one species (as well as how variable that adaptation could be) it would take a human immigrant to highlight the potential and problems of introducing foreign dung beetles to solve the problem of cow dung in the antipodes.

This brings us back to George Bornemissza. When he arrived as an immigrant in Australia, the fly problem had grown so dramatically that no one who visited Australia in the 1950s could be unaware of the swarms of flies that made life for humans and beasts so uncomfortable at certain times of the year. The maths underlying this exploding population was simple. An average adult cow drops 12 three-kilogramme pats of dung every day, and each pat can in weeks yield hundreds of flies (which can lay hundreds more eggs on the freshly dropped dung) generating a perfect storm of flies. If these pats are not disposed of, they will cover a significant percentage of any grazing area. Moreover, an area of rank pasture will grow around the edge of each cow pat. Apart from dung beetles, few creatures enjoy dining close to their excreted waste, and cattle are no different. They tend to avoid rank grass close to their old droppings. The implications of this in Australia, where there was nothing to break down the dung, was that billions of flies were bothering both people and cattle while millions of hectares of grazing land were being despoiled each year.

Bornemissza observed this closely while doing field work in his first position as a graduate assistant in the Zoology Department at the University of Western Australia in October 1951. His research was aimed at reducing populations of buffalo flies, which were causing large losses to farmers, not only by damaging cattle through their biting, but also in wasted and largely unsuccessful attempts to control the problem. He noted the absence of the dung-processing beetles that he recalled from Hungary, and realised that biological control was probably the best solution – it could literally bury the cause of the fly problem.

In December 1954, Bornemissza joined the Canberra-based Commonwealth Scientific and Industrial Research Organisation (CSIRO) as a research scientist in the Division of Entomology. It wasn't until 1965 that what became known as the Australian Dung Beetle Project finally secured funding from the Australian Meat Research Committee and got going under the aegis of the CSIRO. In 1966, a pilot project imported foreign dung beetles into northern Australia to test the prospects for the biological control of dung in the region. Bornemissza travelled to Hawai'i, where he knew several alien dung beetles were already established. He chose species of Hawaiian exotics for shipping to quarantine in Canberra, after which five species were mass-reared for general release. The first species selected for introduction into Australia was the diminutive (12mm) nut-brown *Onthophagus gazella*. This 'tropical strain' from Hawai'i had originally come from Zimbabwe (then Southern Rhodesia) in 1957. In 1970, a strain from the Eastern Cape of South Africa was introduced directly into Australia. These beetles, along with the entomologists, never looked back. Today they inhabit much of the northern-eastern parts of the country, where they are the dominant species in many subtropical areas. Climate-matching models suggest they are likely to infiltrate every corner of Australia except for the dry centre.

The business of distributing the beetles developed into a form of rural direct marketing. CSIRO representatives travelled around Queensland from April 1968 onwards, politely asking farmers if they would allow the release of the dung beetles they had brought with them into the farmer's fields with the purpose of burying their cattle's dung pats. The notoriously suspicious Aussie farmers were surprisingly co-operative. Given the reputation Australian farmers have for directness, the CSIRO crew were surprised by the lack of mockery or derision with which they were received, and in fact received only one complaint from a stud-master, who found the dried and hardened dung pats useful for propping up his irrigation pipes. He, like the others, nevertheless agreed to their experiment and eventually had to resort to more conventional wooden block props for his irrigation system.

This first large-scale release of dung beetles by the CSIRO in 1968 was followed by further releases over the next three summers. In all, a total of 275 000 beetles of four species were let loose, mainly between Broome in Western Australia and Townsville in Queensland. Little *Onthophagus gazella* made extraordinary progress, colonising an area 400 kilometres by 80 kilometres in just two years, and in the process causing spectacular dung dispersal. Two other species also became established. The cattle farmers were so impressed by the results of this trial that they moved to fund a bigger project, based in South Africa, to import a range of dung beetles appropriate for the different climatic regions of Australia.

While the dung beetle headquarters were in Canberra, two small dung beetle research groups were also established to follow the fates of introduced beetles in different climatic areas of Australia. The Rockhampton unit evaluated the effectiveness of the dung beetles in the summer rainfall area, while the Perth unit focused on beetles from a winter-rainfall Mediterranean climate, as well as investigating the ecology of native dung beetles and the bush fly. Another small group was established at Montpellier in Southern France, as a base for the collection of European beetles for Southern Australia, also in the winter rainfall region.

Between 1967 and 1982, the CSIRO imported 55 species of dung beetles for release in Australia. Dr Doug Waterhouse, Chief of CSIRO Entomology between 1961 and 1981, oversaw the introduction of dung beetles from Africa into Australia. Of these, 37 were intended for summer rainfall regions of Northern Australia. Learning from the mistakes of other dung beetle introductions in the world, the CSIRO researchers took pains to match selected beetles with Australia's soil type and climate, and particularly its major rainfall patterns in the cattle-rearing territories. South Africa's summer tropical rainfall region in the northeast, accompanied by the Southern Cape's Mediterranean winter rainfall area, combined with good infrastructure and millions of years of evolution alongside the dung of large mammals (including buffaloes) made the country an obvious choice for dung beetle exploration. Despite the worsening international reputation of the apartheid government, scientists in both countries just got on with the job in pursuit of their goals – dung burial and fly suppression in Australia.

By the 1960s, Australian scientists were well aware of the hidden biological dangers of importing flora and fauna from elsewhere, which meant that a rigorous protocol was put in place to prevent any unwanted organisms arriving with the dung beetles. Adult dung beetles live in and on dung, which translates into a fairly dirty lifestyle. In addition, they carry phoretic mites that nestle in the numerous nooks and crannies of their armoured bodies, hitching a lift between dung pats. To avoid such hitchhikers, only dung beetle eggs were airfreighted to quarantine in Australia. The smooth, globular eggs were harvested from beetles bred in overseas field stations, then packed into sterilised peat moss for transport. These eggs had already been surface-sterilised – washed in formalin supposedly to kill any micro-organisms on their outer surfaces (though the technique has since been shown to fail in this regard).[13] Unpacked in quarantine, the eggs were placed into handmade brood balls for rearing through to adulthood. These emerging adults were then mated and the process repeated, with only their eggs leaving quarantine to be reared to adulthood and finally released in Australia. No wonder some of the imported species never made it out of quarantine.

Others were released in such small numbers that the odds were against them setting up self-sustaining populations – the Allee effect probably did them in. The 53 *Sisyphus mirabilis* released didn't stand much chance, which is a loss for Australians who might never see the spindly beauty of this long-legged, ball-rolling 'spider beetle'. Releasing at least 8 000 individuals, with no less than 500 per site, appeared to be the magic number for success. However, the large European tunneller *Copris hispanus* (named by Linnaeus and investigated by Fabre) seems to have gained a toehold in Western Australia even with just 294 individuals released. Brood care, where the adult beetles stay underground with their precious brood balls until the new adults emerge, may well have improved the chances of successful reproduction of the founding group. As a spring-active, winter-rainfall species dominant in its home range, *Copris hispanus* holds the promise of doing damage to bush fly populations that build up before the African-origin dung beetles have woken up from their winter diapause (a resting stage similar to hibernation). Only 2 105 *Copris elphenor* (of African origin) were released between 1978 and 1983, all at one site in Queensland. This brood-caring tunneller is still hanging in there, and has the potential eventually to reduce buffalo fly numbers in that region.

With the release of more than 1.7 million dung beetles across the continent, the project has yielded impressive results. In areas with two or more species of introduced dung beetles, the CSIRO found that bush fly numbers were reduced by 88%, and the survival of a bush fly egg to adult stage was reduced by 99%. However, the funding for the Australian Dung Beetle Project was withdrawn in 1985 after the restructuring of the Australian Meat Research Committee.

Over the twenty-year life of the project, the impact of the beetle introductions went far beyond the initial focus on managing the buffalo fly problem. They were found to have had a positive impact on soil, water and pasture health. The burrowing dung beetles not only buried the nutrient-rich dung and aerated the soil, but improved access for another great friend of good soil, the earthworm. These take over after the beetles have finished their burrowing and breeding cycle. The runoff from fresh dung into water has also diminished, with all the attendant advantages of less polluted water.

In 1994, the Double Helix club of the CSIRO introduced the Dung Beetle Crusade, a brilliant way to engage children with nature and teach them practical scientific skills. Some 1 300 schoolchildren across Australia were tasked with collecting dung beetles in their local area. As a result of these collection efforts, dung beetle researcher Penny Edwards was able to map the surviving introduced species and propose that at least 26 of the 50 species that had been introduced had survived and established self-sustaining populations. Subsequent investigations by Edwards in 2007 concluded that 23 of the 50 species of dung beetle introduced by Bornemissza and his successors were still established and thriving all over Australia.

Despite the accomplishments of the established species, it was clear by the late 1990s that more dung beetle species were needed. Even though the main project had closed in 1986, scientists continued to review the ongoing spread and impact of the imported beetles. Edwards's 2007 study found there were notable gaps in the network and distribution of dung beetles. Australia's tropical and sub-tropical cattle grazing areas are currently served by seven to 13 species of dung-burying beetles, but temperate pastures have no more than four or five species (by contrast, many African habitats can boast up to 100 species in a single dung pat).

Across the winter rainfall region of Southern Australia, most of the introduced species emerge in late spring and reach peak numbers during summer, with only one species active during the autumn-winter period. That early spring gap in dung beetle activity coincides with a spring influx of migrating bush flies. As a result of this significant hiatus, when the flies can breed unhindered by their main competitors in the dung business, a group of mainly retired dung beetle researchers identified two more European species for introduction. These 'Old and Bold' retirees of the Dung Beetles for Landcare Farming Committee used their cumulative decades of experience to further refine the Australian dung beetle programme years after its official closure, revealing their commitment to the conviction that dung beetles should be part of Australia's agricultural toolbox. *Onthophagus vacca* is a tunneller in which both sexes brood care, and *Bubas bubalus*, a dusk-active tunneller and brood carer: both are active in the spring. The Department of the Environment approved the additions, and the first newly emerged adults of *Onthophagus vacca* and *Bubas bubalus* were collected in Southern France in the spring of 2012, 2013 and 2014.

The new method of introduction had one radical variation from previous efforts. Instead of importing eggs, adult beetles were cleaned and air-freighted to Australia where they were mated in the high-security CSIRO Black Mountain Containment Facility in Canberra. Both species selected for introduction produce only one generation each year, and because they are active in the Northern hemisphere spring, they were six months out of synchrony with the Southern hemisphere seasons. However, many dung beetle adults and larvae go into a resting stage over winter known as diapause that allows these insects to survive unfavourable conditions. This is triggered by environmental cues, such as day length and temperature. The new beetles had their seasonal clocks reset by researchers who used high temperatures followed by a short, simulated winter to trick the beetles' physiology into releasing them early from diapause. This novel solution also allowed *Onthophagus vacca* to complete two breeding generations in one year.

Farmers who are prepared to work with the introduced dung beetles have to think through their relationship with their six-legged stock. They have to limit the parasiticides and dips known to have a negative impact on dung beetle survival. Just as they cater for the rest of their animals, farmers must ensure that they have grazers active in the paddocks in which the dung beetles are to be released. It all seems simple, but even farmers need a surprising amount of education on insects, particularly ones they are unfamiliar with. Dung beetle distribution is now a private enterprise in Australia. There is a reassuring resurgence of dung beetle popularity among cattle farmers in Australia, who understand that the value of the beetles extends far beyond limiting fly breeding sites. In a population of 26 million people, Australian farmers run 28 million cattle across 200 million hectares of land. The role of dung beetles in this enterprise is now significant, and the Australian government in 2018 committed twenty million dollars to further research into the effects of dung beetles on soil health.

Considering that biological control is often the Cinderella branch of pest control, the success story of dung beetles in Australia and, to a lesser extent Hawai'i, is a measure of the potential for the effective use of beneficial insect control. Dung beetles are particularly remarkable in this context because there is as yet no evidence that they alter an environment in any negative way. There are reports of some of the more vigorous species like *Onthophagus gazella* and *Euoniticellus intermedius* displacing indigenous species in the US, but this has yet to be rigorously proven. Dung beetles untiringly ply their grubby trade wherever they find themselves, even if they are not a one-size-fits-all solution. And while they have fallen from their dizzying heights as gods of human resurrection, they do in fact resurrect damaged soil. They have undoubtedly made Australia a far more liveable environment, making it probably the first country in the world where an insect has transformed human cultural practices. Australians can now enjoy a vibrant outdoor pavement-café culture, and the Australian salute of 'G'day Bruce' accompanied by a sweep of the hand across the face may be a greeting doomed to extinction.

CHAPTER FIVE

Of elephants and dung beetles

THE AUSTRALIAN DUNG BEETLE story is largely attributable to a successfully orchestrated and thoroughly researched biological solution to an imported problem. It was Australia's good fortune that George Bornemissza arrived from Hungary when he did, and was able to prove the practicality and efficacy of dung beetles. He ran against the trend of the post-war tendency to reach for quick, cheap and apparently effective solutions. Dung beetles, despite proving themselves to be relatively adaptable little transformers of the earth, were unable to keep up with the excitement of a world in which it was possible simply to sprinkle an almost magical dust known as DDT (dichloro-diphenyl-trichloroethane) to end the life of an unwanted insect. With a limited understanding of how the natural world functioned, most insects were seen as problematic; differentiating between beneficial insects and harmful ones was not particularly high on the list of concerns of farmers driven to produce ever-increasing volumes of food.

The world at the end of the Second World War was a dismal time for insects: the myriad species of useful insects suffered as much as the few detrimental ones because DDT and the other new poisons did not differentiate between them. DDT is a member of the chlorinated hydrocarbon group. It was first synthesized in 1874, but its effectiveness as a pesticide was discovered only in 1939. During the Second World War, it was used for everything from delousing prisoners to controlling malaria.[1] It was inevitable that following that war, DDT was regarded as the new wonder solution to insect pests. It was cheap, versatile, initially very effective and its residues kept the pests away for an extended period. Looking at only one country's consumption of DDT gives an idea of the scale of its usage. Up until 1972, approximately 1 350 000 000 pounds (or 675 000 tons) of DDT were used in the US, making it one of the most widely used pesticides in that country.

Biological control compared to DDT was simply too slow, but then an odd development gave the users of DDT and other pesticides pause. Farmers started to notice a resurgence in the numbers of pests where they had been spraying, accompanied by the worrying appearance of new pests that had never previously been present. Even worse, relatively benign insects which had only caused trivial damage in the past, now became major pests in spite of almost weekly spray regimes. It seemed as if the pests were thriving and getting stronger on the poisons supposed to kill them. The crops had weakened, but the insects had become resistant to DDT and were multiplying everywhere. All of the chemicals were failing. Azodrin had gone from having 'solved' the problem of bollworms on cotton crops in the 1950s to apparently becoming a stimulus to the bollworms. The same effect was found across a spectrum of toxic chemicals and their target insects.

What was most remarkable was the relatively short time span it took for insects to adapt to DDT. The long-term ineffectiveness of DDT meant that after 1959, its usage in the US declined greatly (dropping from a peak of approximately 80 million pounds in that year, to just under 12 million pounds in the early 1970s). Over eighty per cent of these pesticides used between 1970 and 1972 were applied to cotton crops, with the remainder being used predominantly on peanuts and soybeans.

The decline in DDT usage was the result of increased insect resistance and the development of more effective alternative pesticides that used different compounds and different target sites in the insect's physiology. These factors, along with a growing public concern over adverse environmental side effects, led to increasing government restrictions on DDT use. Nevertheless, DDT was still used both domestically and abroad. Large quantities of DDT had been purchased by the US Agency for International Development (USAID) and the United Nations and exported for malaria control. DDT exports increased from 12 per cent of total production in 1950 to 67 per cent in 1969. However, by the 1970s, exports also started to decrease.

The publication of *Silent Spring* by the marine biologist Rachel Carson in 1962 drew attention to the damage that had been done to the environment as a result of the indiscriminate broad-scale usage of toxic pesticides. It was the first popular book that described a view of the world as an integrated whole. It clearly spelled out that mechanical solutions used in isolation had caused massive and unforeseen environmental problems that were still not fully understood. Carson's book, which sold over two million copies, had a profound impact on the nascent environmental movement of the 1960s. She drew attention to the chain of nature, showing how once a bird or fish ate an insect poisoned by DDT, the poison would eventually end up in the food chains of which humans were often part. It explained how DDT came to be found in human breast milk, something no parent would consciously sanction. Although Carson was certainly not the first to view the world in this way, she stood at the cusp of a new emphasis on and engagement with the environment. Three hundred years earlier, Maria Sibylla Merian had been one of the first to paint a world that reflected the connectedness of flora and fauna; however, the number of scientists, artists and naturalists exploring that view had always been small (even after Darwin connected the dots in proposing his all-encompassing theory of evolution).

Whereas the problems for agriculture in North America and Australia were a product of rapid ecological change brought about by new crops and farming methods, changes in land usage in other former and existing European colonies were of a different nature. Africa, the home of more species of dung beetles than any other continent, owed this variety to the presence of its iconic wildlife. Dung beetles had evolved a complex and rich relationship with the animals of Africa, largely independent of interference by modern humans and their version of agriculture. It was in fact the perceived threats to wildlife from burgeoning populations from the mid-1950s onwards, particularly in the soon-to-be-independent colonial territories, which prompted the very first studies on the interaction of dung beetles with native mammals in the wild. It was dung beetle research in Africa that revealed the complex and infinite variation in their adaptations to different environments and food opportunities, and which was to eventually elevate them into the pin-up creatures of evolutionary studies.

The recent history of wildlife in Africa (particularly that of the large species) reveals a high level of exploitation of wild animals as a resource long before the era of European colonisation. Pointing the telescope at elephants – a source of abundant food for dung beetles – illuminates the complex interweave of the continent with its wildlife.

The concept of Africa as an undisturbed Eden prior to the advent of Western colonialism is thoroughly inaccurate. For millennia the natural resources of the continent (both animal and human) had been exploited, with internal slavery common. Rhino horn had been traded both within and beyond the continent for centuries (as had ivory), along with live animals and their skins. Arab traders and local rulers along the East coast of Africa had a long history of monopoly and control of a lucrative and destructive trade, where ivory derived from the slaughter of elephants was a key commodity.

Along those trade routes, thousands of African slaves were brought from the interior of Africa to the coast for further shipment to the east, where they were exploited in a variety of ways. The major difference between the East and West coast slave trade was the gender of those sold into slavery. On the West coast the preference was for men (who were sent to plantations in the Americas) while on the East coast women were more in demand. Male slaves on the East coast, although exported in smaller numbers than female slaves, were used to carry the heavy ivory to the coast. Once there, some of these slaves were sold while others were used to bring goods (purchased with the proceeds from the sale of the ivory) back to the interior. When the European nations colonised Africa, there was no primal state that lay waiting for them to impose whatever rules they cared to enforce. The continent already had its own complex social, political and economic structures, many of which were in a state of flux easily destabilised by the new invaders with their superior weapons. Power and wealth were therefore frequently linked to the exploitation of the resident wildlife.

Ivory had been used for centuries for everything from thrones to combs, but we have no adequate records to account fully for the extent of the trade, other than the knowledge that trading routes were constantly being extended further into the African interior as populations of elephants were depleted. Nevertheless, when the industrial revolution began in Great Britain and spread beyond it, the trade (particularly in ivory) received a radical boost. However, as we have no figures for the quantities of ivory exported prior to colonisation, we cannot be certain whether or not these levels of exploitation had been going on for centuries.

The rising affluence of the Victorians saw ivory used for everything from cutlery handles, door handles, combs, pistol grips, drawer pulls and buttons, to letter openers. However, it was primarily the leisure pursuits of the middle classes – music and billiards, in particular – that drove the demand.

Pianos were a symbol of social mobility in the late 1800s, and between 1850 and 1914 global production of pianos increased from 43 000 in 1850 to 600 000 in 1910. Ivory was the favoured material for keys, because it was neither too cold nor too hot, and with its minute porosity, it had the ability to absorb perspiration. It also was reputed to have an agreeable feeling for the piano player. The US was using nearly two hundred tons of ivory per annum by 1913 for keyboards alone. This figure did not take into account the other huge consumer of ivory: the manufacture of billiard balls. In 1922 T. B. Wadleigh, secretary of the Illinois Billiard Association, commented that he was safe in saying that 4 000 elephants were killed every year to supply the ivory departments of billiard makers in the US alone.[2] What made the use of ivory for billiard balls even more destructive to elephant populations was that the manufacturers placed a premium on small tusks – 'scrivelloes' – particularly those from females, which are straighter than those from males.

We now know that elephant herds are primarily matriarchal. Killing off females in the quantities demanded by the recreational activities of the Victorians represents an unrecorded environmental and ecological impact on one keystone species in Africa about which we know very little. Matriarchs are the repositories of wisdom in elephant families, holding memories of local resources, migration routes and controlling the behaviour of their group. Losing them is like losing the head of a family. No one at the time knew that elephants had such a sophisticated matriarchal society, and even if they had, it is debatable whether they would have cared.

Ironically, one of the strange by-products of the decimation of wildlife in Africa was the emergence of the first conservation initiatives, frequently from the very people who (having done so much of the slaughtering) had seen first-hand evidence of the result of their sport. The rise of the radio fortunately ended the boom in pianos in the 1920s and the demand for ivory lessened, creating a brief breathing space for the surviving elephants. However, along with other large herbivores, elephants were facing new challenges as human population densities increased and less land was available for the animals. The equivalent of the Columbian exchange in Africa resulted in the arrival in the 1800s of diseases new to the continent.[3] Just as traders returned from the Americas with potatoes, maize and tomatoes (having swopped them for measles and smallpox), increasing trade into Africa by other European colonies brought with it unwanted impacts on livestock and indigenous wildlife.

One of the most devastating new arrivals was an infectious disease of cattle known as the rinderpest. *Rinderpest morbillivirus* was carried into Africa initially via Egypt by a few head of cattle that arrived on a ship at the port of Alexandria in 1841. Some of the infected animals were sold, and the virus (which is related to the measles and canine distemper virus) spread with astonishing speed and devastating results. Approximately 665 000 domestic animals died in Egypt alone. The disease was believed to have died out by 1843, but it returned via cattle from Southern Egypt shortly thereafter, and led to the loss of about 90 per cent of Egypt's cattle.

However, the subsequent great African pandemic of rinderpest did not originate in Egypt. It swept through the continent from 1889–1897, having its possible origin in Ethiopia, which was invaded by an Italian expeditionary force that arrived at the port of Massawa in November 1887. The force came with cattle from India which carried the virus, and its spread through Ethiopia was rapid and deadly. Rinderpest then proceeded to move through Somalia into Uganda and West Africa, and south to Kenya, Tanzania, Malawi and Zambia. Everywhere it spread, it caused massive mortality among both immunologically naïve domestic and wild animals, with some herds losing up to 100 per cent of their members. The wildlife affected were the even-toed antelopes, including buffalo, eland, kudu, wildebeest and other ungulates. When the rinderpest reached Zambia, the Zambesi River put a temporary halt to its spread, but by 1896 it had crossed the river and started its deadly journey into Southern Africa. Better facilities for treating infected animals in South Africa and for preventing the spread of the disease meant that 1901 saw the last serious outbreak in South Africa. Meanwhile, the loss of livestock in the many cattle-based cultures of Africa had a dramatic social and ecological impact, leaving large tracts of formerly grazed land empty of livestock and the populations dependent on them.

The rinderpest created a vacuum into which the tsetse fly moved. This serious pest fly is not one that any dung beetle could have an impact on, because the maggot has no need for dung; it develops in splendid isolation within the genital tract of its mother, nourished by a 'milk gland'. This blood-sucking female fly gives birth to a fully formed larva that immediately burrows into the ground where it pupates. When the rinderpest reduced the wildlife numbers, particularly of some ungulate species, humans became the next best source of blood for the tsetse fly, and the incidences of sleeping sickness increased.

Both male and female tsetse flies take blood meals from mammalian hosts. In the process, infected flies transmit a protozoan micro-organism (similar to the malaria parasite) which causes *Trypanosomiasis*, commonly known as sleeping sickness. The colonial governments of the time decided that, where possible, it was safest to remove whole populations from areas affected by sleeping sickness. This mass movement of large groups of people frequently led to conflict with both resident populations in the areas to which the displaced people were relocated, and with wildlife that had their migration routes impacted. For example, an outbreak of sleeping sickness at the turn of the twentieth century in the region of North Bunyoro in Uganda resulted in an enforced evacuation of the entire population to Central and Southern Bunyoro. The area from which they were removed was then declared a game reserve, but the relocation areas experienced the pressure of increased human population density, which caused subsequent conflict between humans and elephants whose range overlapped the newly settled area.

The conflict was so severe that it resulted in the establishment of a game control department, whose policy of harassment and confinement resulted in the shooting of at least 15 000 elephants between the years 1925 and 1965. Game parks established as a consequence of the complex pattern of the disease's outbreaks were frequently located in areas that had been made uninhabitable largely by disease, and which were not defined by animal use or abundance. No research or observation on the patterns of animal migration, or their usage of flora and other fauna, was conducted. Rather, there was a perception that animals were everywhere, so they could quite easily be confined to areas where humans found it difficult to survive as a result of factors ranging from inadequate water supply to the presence of malaria and other diseases.

The containment of some species of animals in areas they might only have migrated through in previous times presented a whole new dimension to the concept of adaptation and survival. Wild animals, in addition to being confined to game parks, became dependent on the management decisions of humans. Corralling and limiting the movement of wild animals eventually stimulated a new focus in the biological sciences which led to the eventual development of ecological studies. This new biological discipline emerged from the need to understand managed and transformed landscapes with a high level of human interference, and led to another novel profession: wildlife management. Such a career was a daunting one in the mid-twentieth century, when relevant information was virtually non-existent.

The complex eco-systems to which the first generation of conservation ecologists were introduced in the 1960s was one of which they had only limited knowledge. For instance, they did not know that elephant society was matriarchal; they believed that the large bull males led the groups. Not only did they know little, they had to adapt as they went along without the benefit of many twenty-first century observation techniques and recording technologies (like radio collars) which are almost standard in today's population studies.

Nevertheless, these newly minted conservation ecologists were sometimes arrogant about their science-based decisions, which meant that they frequently clashed with the wardens and colonial naturalists who had not that long before established the wildlife parks. All of these practitioners were becoming aware that they were in some senses custodians of 'their' animals, as well as the whole planet. While this realisation may have conferred an awe-inducing power on them, it sometimes led to a seductive and dangerous god-complex.

The concept of wildlife parks had been sold to the soon-to-be independent former colonial territories not just as wildlife sanctuaries, but also as sources of tourist revenue. It was hard to reconcile this with the demands of young scientists who often felt they had superior techniques for wildlife and reserve management – if only they could be left alone to do their work. Science had come a long way from simply naming and ordering the abundance of the world. It was now easing into a new era built on an acceptance of Darwin's theory of evolution, along with other advances in understanding biology. The dispute over creation was no longer central to biology: evolutionary studies had revealed there was no need to prove anything about a theoretical or real Eden. Now the last 'Edens' were being fenced and managed, and scientists were called upon to do the necessary research to preserve them and their contents.

The people erecting and managing the parks were as guilty as the broader public of an obsession with the mega-fauna and the romance of Africa, which had portrayed the continent as a primal landscape teeming with wildlife. No one except for our hardy seventeenth and eighteenth century collectors, who supplied dung beetles among other non-vertebrates for the busy natural accountants in the northern hemisphere, had bothered to look at the composite systems within which animals lived. Dung beetles were overlooked along with the myriad other creatures that failed to satisfy or did not fit the image of Africa's majestic wilderness. It is almost impossible to find a mention of dung beetles in early wildlife books about Africa, or any other creature not intended for the table or the trophy wall.

The dung beetle features occasionally in African tales. The Kamba people of Kenya, for example, cast the dung beetle as a cursed individual. Cursed by his dying mother for refusing to prepare her food because he was too busy making bricks for his house, he is now doomed to roll his dung forever, but never to build a house. This Sisyphean tale, similar to the metaphorical tales about dung beetles from ancient times, reveals no innate understanding of the value of the beetle to the world in which it lived. This role took a very long time to be understood. The very earliest paintings in Africa South of the Sahara show no dung beetles. The San people noticed and recorded termites and bees, but not dung beetles. Maybe only settled pastoralists stayed in specific spots long enough to follow and decipher the annual cycles of the beetles, then incorporated these cycles into their mythology or copied their manuring techniques.

The Africa portrayed by early explorers and collectors was never static or pristine. The history of both its human and animal populations (looking at accounts of slavery and elephant exploitation) prove that changes in population dynamics of humans and animals were never at a point of primal equilibrium that we know of within historic times. We have at best an imagined idea of what Africa might have been like during the Western Middle Ages. However, in the same way that we do not know how climate change drove wildlife from the Sahel area millennia ago, we simply do not know enough to do more than draw very broad brushstrokes in depicting how Africa should ideally look in terms of environmental balance. Early conservationists were nevertheless trying to do precisely that. They wanted to undo the outcome of recent human depredations, but they were working with an already fundamentally changed landscape. Africa was going through new changes that were at odds with attempts to set the clock back to some imaginary pristine time. Movements towards political independence and economic and social change were gathering momentum. The national parks were seen as but small bulwarks against these inevitable changes.

Although Virunga National Park (previously known as Albert National Park) in the Belgian Congo is cited as the oldest national park in Africa (having been proclaimed in 1925), South Africa was the first place to create game reserves. The Hluhluwe and Umfolozi reserves were proclaimed in 1895, preceding the Sabie Game Reserve in 1898 (which became the Kruger National Park in 1926). This area of 19 633km² was the legacy of Paul Kruger, President of the then

South African Republic, and represented a desire to limit the loss of animals for sport hunting. Initially, the park was dedicated to protection of 'game', and tourists did not visit because there were no facilities for them. Conservation in those days consisted of simply leaving what was believed to be wilderness, as wilderness. The plan in its most basic form was to proclaim an unfenced area as a game reserve and to shoot carnivores that might compete for game. Tourism started when the National Park was proclaimed, stimulating the construction of roads and rest camps.

In Kenya, the Nairobi National Park was proclaimed in 1946, followed by Tsavo East Park in 1948. Hwange in Southern Rhodesia had been a game reserve since 1928, but only became a National Park in 1961. Tanganyika's Serengeti National Park was established in 1951, and in Uganda the Murchison Falls and Queen Elizabeth National Parks were both formed in 1952. These artificial entities (frequently located in areas where humans had lived but had sometimes been forced out by resurgent sleeping sickness) were created with limited budgets, dedicated naturalists, and a few very able and competent people. These areas initially formed unfenced 'islands' within which nature could apparently revert to the imagined pristine, pre-human world: a further illusion, as in many instances they were places where indigenous populations had found ways to survive alongside the wildlife. The concepts were both flawed and difficult to execute, but they represented the best hope for the diminished populations of wildlife unable to compete with the growth in the human population and their livestock.

Tsavo National Park in Kenya is of significance to dung beetles and the way they have changed our understanding of the biological world because it was the first place the beetles were studied in relation to the animals on whose dung they depend and which they recycle. This was a move away from the single-species approach to ecology, and it signalled the start of a new era in our understanding of the dynamics of ecosystems.

1
Ulisse Aldrovandi, who (in the
sixteenth century) became the first man
to describe the underground brood
chamber of a dung beetle. Lithograph
by Pierre Roch Vignéron.
Courtesy of Wikimedia Commons.

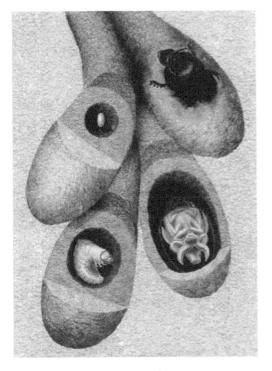

2
A hypothetical scenario, which
illustrates the life cycle of a dung beetle
as it passes through each life stage in a
buried brood ball, and is now ready to
emerge as an adult beetle.
Courtesy of Utako Kikutani.

3

Mark Catesby's tumble turds from Carolina, with *Lilium* sp *martagon canadense*. Catesby recorded the flora and fauna of the South-eastern US and the Caribbean in the eighteenth century. *Courtesy of the Smithsonian Libraries and Biodiversity Heritage Library.*

4

The Good Scarabaeus. Albrecht Dürer's *Stagbeetle*, 1505. Originally a dung beetle in the fourth century, it had metamorphosed by the fifteenth century into a handsome male stag beetle. *Courtesy of Getty's Open Content programme.*

2 mm

5

The beautiful *Oniticellus formosus* is a dung-dwelling species, producing its brood balls inside a dung pat (where the female stays, adding dung to the ball, for more than a week). The species colonises older dung, thus avoiding being disturbed by large ball-rollers.

Courtesy of the Scarab Research Group, University of Pretoria.

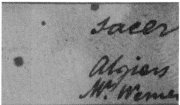

6a and b

Scarabaeus sacer, the sacred scarab. A pinned specimen from Linnaeus'
collection, presumably identified by him because he was the first person to
formally describe and name this species. The accompanying collection label
indicates that it was collected in Algiers, North Africa, by a Mr Werner.
Courtesy of the Linnean Society of London.

7

Pataikos was a protective Egyptian deity associated with Horus through the scarab, Khepri, astride his cap.

Courtesy of Brooklyn Museum, Charles Edwin Wilbour Fund.

8

Two genetically identical Agouti mice that look very different because of epigenetics in their development. When the pregnant yellow mother was fed a diet rich in nutrients (such as folic acid and vitamin B12) the agouti gene was switched off in the pups which are, consequently, brown and thin; not fat and yellow.

Courtesy of Randy Jurtle and Dana Dolinoy.

9

The cover of Thomas Moffet's *Theatre of Insects*, published in London in 1634. Note the dung beetle heading down the left hand side of the page.

Courtesy of Wikimedia Commons.

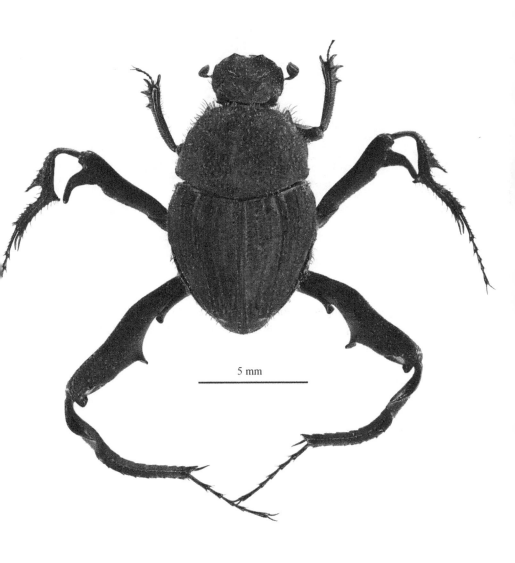

5 mm

10

A male *Sisyphus mirabilis*, also referred to as a 'spider dung beetle'. A particularly elegant species in the genus *Sisyphus*, named after the Greek king who was punished by the gods by having to push a huge rock repeatedly to the top of a hill, from where it would roll back down to the bottom. There are more than 40 species of spider dung beetles in this genus.

Courtesy of the Scarab Research Group, University of Pretoria.

11

A mass of one species of dung beetle, *Pachylomerus femoralis*, on a rhino dropping in South Africa. Such intense competition is thought to have driven the evolution of ball rolling behaviour in dung beetles. Up to 100 species of dung beetles can occur on one dung pat and number 1000s of individuals.

Courtesy of Costas Zachariades.

12

The dung beetle dance. Climbing to the top of the ball, the beetle circles and scans the sky to orientate itself relative to the sun and other cues in the sky.

Courtesy of Marcus Byrne.

13

Wearing a cap prevents this beetle seeing orientation cues in the sky. As a consequence it rolls its ball around in circles, like a human lost in a featureless desert.

Courtesy of Marcus Byrne.

14

Kheper lamarcki wearing silicon boots to protect its feet from the hot soil. Beetles without boots climb onto their ball more often to cool down, indicating that the ball has been incorporated into its thermoregulation behaviour, in addition to its roles in feeding and reproduction.
Courtesy of Adrian Bailey / baileyphotos.com.

15

A thermal image of a dung beetle escaping the torment of walking on hot sand by perching on top of its cool ball. Pale, yellower colours indicate hot areas, while dark, bluer regions are cooler.
Courtesy of Jochen Smolka.

16

A dung beetle wearing silicon boots in an experiment to test the effect of hot sand on its ball rolling and thermoregulatory behaviour.

Courtesy of Marcus Byrne.

17

A major male *Proagoderus watanabei* with impressive horns for fighting other males.
Courtesy of Rob Knell.

18

Dung beetles facing off for a fight. Even though both possess a ball, they will still engage in a brawl with any other beetle that gets close enough to grapple with it.

Courtesy of Chris Collingridge and Lund University.

19

A series of *Proagoderus lanista* females (left column) and males (right column) showing the influence of body size on male horn size: large horned major males are at the top; small minor males at the bottom look like females.

Courtesy of Douglas Emlen.

20

A species of *Pachysoma* galloping across the sands of Namaqualand. This singular behaviour is only seen in three species of this beetle, and in no other insect.

Courtesy of Jochen Smolka.

21

Kheper lamarcki rolling at speed. Head down and moving backwards, the small eyes at the back of the head continually track cues in the sky to maintain a straight rolling path.

Courtesy of Chris Collingridge and Lund University.

22

Press photo of Marcus Byrne from *The Star* newspaper, accompanying the news that dung beetles can orientate by the stars of the Milky Way.

Courtesy of Chris Collingridge.

23

Researcher Basil el Jundi noting the response of an orientating dung beetle to a switch in the intensity gradient of the sky, caused by the use of an overhead filter, which triggers the beetle to reverse direction.

Courtesy of Chris Collingridge.

24
Researchers Marie Dacke and James Foster testing the orientation responses of nocturnal dung beetles under different night skies in South Africa. Insects can't see red light.

Courtesy of Chris Collingridge and Lund University.

25

A diurnal dung beetle, *Kheper lamarcki*, responding to a green LED light spot in an indoor arena.
This allows precise manipulation of elements of the sky, which is not possible out of doors.

Courtesy of Chris Collingridge and Lund University.

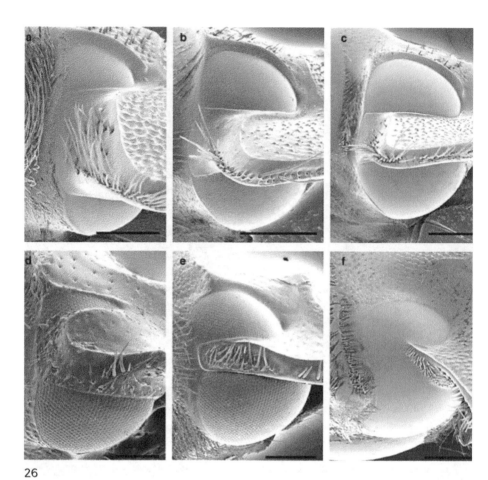

26

The range of eye size and shape in dung beetles. The top row are ball-rollers, the bottom are tunnellers. The lefthand column are diurnal and the righthand nocturnal. Crepuscular (dawn and dusk active) species are in the middle column.

Courtesy of Marie Dacke and Lund University.

27

Researchers engrossed in dung beetle orientation experiments at their field station in South Africa. Having finished with diurnal beetles during the day, they will set similar tasks for the nocturnal species that appear at night.

Courtesy of Chris Collingridge and Lund University.

28

A dung beetle orientating as it rolls its ball. Alongside is a prototype robot, which has been designed to use the same cues in the sky as the beetle to track along a straight path.

Courtesy of Marcus Byrne.

29

A dung beetle trying to roll a fake dung ball, used in an experiment to test how dung beetles distinguish a dung ball from other objects.

Courtesy of Marcus Byrne.

30

Digitonthophagus gazella. The little insect that is changing the world as the most widely distributed dung beetle. Mostly deliberately, sometimes accidentally, it has found its way from Africa to ten countries around the world.

Courtesy of Shaun Forgie and Dung Beetle Innovations.

31

Euoniticellus intermedius, hornless female and horned male, both about one centimetre long. Introduced to Hawai'i, mainland North America and Australia, this small tunnelling species competes for dung with dung-breeding pest flies.

Courtesy of Marcus Byrne.

32

Pachysoma cancer. Known for over a century only from museum specimens, this large, enigmatic flightless species was found surviving on land-mined dune fields in Angola. Its rediscovery in 2015 has resulted in its demise. Now collectors can sell a single, pinned specimen for 1500 euros.

Courtesy of Birgit Rhode, Landcare Research and Shaun Forgie, Dung Beetle Innovations, New Zealand.

33

Kheper nigroaeneus. Named after the Egyptian god Khepri, this magnificent ball-rolling species occurs over much of Southern Africa.

Courtesy of Chris Collingridge.

34

George Bornemissza with his beloved beetles. He was the instigator of the Australian dung beetle programme, which released over 1.7 million beetles onto the continent to control dung-breeding flies. *Courtesy of CSIRO Australia.*

35

Jean-Henri Fabre discovered a great deal through careful observation of dung beetles in the meadows of Southern France.

Courtesy of Wikimedia Commons.

36

Doug Waterhouse, CSIRO Entomology chief, looking nonplussed in a publicity shot showing how handmade brood balls were used for introducing dung beetles into Australia.

Courtesy of CSIRO Australia.

The study came about largely because of a dispute over the impact of elephants on the environment. Confining large numbers of elephants into one defined park had resulted in what appeared to be a great deal of environmental damage to the trees and other vegetation. This had been observed earlier in Uganda at Murchison Falls; it also became apparent in Tsavo during a severe drought, alarming park staff who could see the impact of a large ecosystem engineer (the elephant) in a restricted space. Based on the Ugandan observations, scientists felt compelled to act and there was a fierce public debate as to how the problem should be solved. This resulted in the first scientific study of elephants in the park, which unfortunately started a dispute between the new breed of ecologists and the resident wardens. The latter were knowledgeable and emotional about the animals they cared for. This was particularly true of David Sheldrick, who had worked hard to save the remnant populations of Kenya's elephants from poaching in the giant 20 850km² Tsavo National Park during the 1950s. Within a year, anti-poaching teams of former soldiers and trackers (who were either ex-poachers or people recruited from areas far beyond the borders of the park) had put a stop to the problem. As outsiders, the latter had few local social networks and so were not bound by ties of kinship with or social obligation to the people living around the park. So it was not completely unexpected that when Richard Laws (a Cambridge large mammal ecologist appointed as research director at Tsavo in the early 1960s) demanded the culling of 300 elephants for research purposes, serious tensions arose.

The clash was unfortunate, and one outcome of the culling (that highlighted the problems with this method of research) was the sad discovery that one of the groups culled had been led by a blind matriarch who had somehow managed to lead her group safely through the years of excessive poaching, only to be culled. Although Laws's research revealed a great deal of hitherto unknown information, mainly on the reproductive biology of elephants, he and Sheldrick held opposing views on the ethics of killing elephants. Sheldrick advocated allowing nature to follow its course during the drought, while Laws held that timely interference was appropriate. Laws's claim that mass culling the way that he did it did not affect other elephant populations is now judged (unfairly) with the hindsight of modern knowledge. He had no way of knowing at the time that elephants can communicate over large distances through very low frequency infrasound inaudible to human ears. So his claim that killing certain herds would not be known to other herds is now interpreted as scientific hubris and ignorance.

The quarrel over culling came to a head when Laws announced his desire to kill a great many more elephants for his research. At that point, he was asked to leave his position. Laws was considered to be indifferent and dismissive of the wardens' work and research, and somewhat disdainful of the value of years of practical, on-the-ground experience. Sheldrick had indeed conducted research on elephants in their environment. Daphne Sheldrick, in her Tsavo blog, describes how David had undertaken experiments on orphaned elephants to understand, for instance, how long it took an orange to pass through an elephant's gut and appear in its stool. He also weighed their dung against an estimate of fodder intake and analysed the protein content of the dung to calculate the elephants' digestive efficiency. Nevertheless, similar experiments had already been conducted with zoo animals.

Despite this inauspicious start to the presence of scientists at Tsavo, the realisation that elephants and dung beetles might have a relationship with one another encouraged other scientists to examine different aspects of the park's ecology. A Canadian biologist, Dr John Goddard, studied rhinos in Tsavo from 1967–68. He reported that there were between 30 000 and 40 000 elephants in an area of 44 000km². The damage to the woodland trees, he noted, was phenomenal. The elephant-assisted demolition of this vegetation during the subsequent drought led to the demise of the black rhinos. This loss at least highlighted the interconnectedness of the organisms in the ecosystem.

Malcolm Coe arrived at Tsavo from Oxford University as a perceptive young naturalist from a country where insects were the subject of much frontline ecological research. Understandably, he latched on to the dung beetle–elephant interaction among the other interests he pursued in the park. Although we know that agricultural studies had been conducted on dung beetles' ability to recycle cow dung, this was the first time a naturally evolved relationship was examined in the wild. In addition, Coe developed a scientific respect for dung beetles, and went on to spread this interest, influencing new research in the field after he returned to Oxford. Coe and his fellows were fascinated by the incredible recycling job that the Tsavo dung beetles performed, and passed on this passion in their writing and popular lectures presented to the Kenyan public on the subject.

Coe calculated that the Tsavo elephants defecated up to 17 times a day. At that time, the density of elephants was approximately 4,4 elephants per square mile, which translated into about 75 piles of dung a day, and therefore 27 000 piles of elephant dung being dropped on each square mile of land per year.[4] Coe must have wondered why every living thing wasn't tripping over great piles of elephant droppings, and credited their absence to the dung beetles. This is one of the first calculations we have of the real significance of dung beetles in the wild and their importance to an ecosystem – in this case, helping elephants to unknowingly distribute nutrients across a landscape and into the soil. Coe's calculations highlighted a key relationship between a huge mammal and a small insect. In the absence of dung beetles, termites did the job. But dung beetles did it differently: their excavations are opportunistic (they dig where the dung drops) and ephemeral (they dwell in a given burrow for no more than a year at most). In addition, they are important distributors of the seeds found in the dung.

Dung beetles are primarily seen in the rainy months, while termites take over most animal waste recycling in the drier months. Dung beetles are therefore active at a time when many wild animals are usually producing their young, and it is they who keep the soil healthy and its surface clear, suppressing filth fly populations in the process.

The importance of understanding this role of dung beetles in the wild, as opposed to those in Australia and Hawai'i, lay in the fact that for the first time ecological studies of natural ecosystems were being conducted. Not only were there niches in nature that complemented each other and worked in some sort of harmony, but this interaction was dynamic. If we consider how many elephants have been killed through the centuries, we can only begin to guess at the impact this might have had on dung beetle evolution and adaptation. Dung beetle diversity drops whenever people move in, replacing ecosystems with agricultural alternatives, even when cattle grazing is the replacement for the wilderness. We will literally never know how dung beetle history has been influenced by recent transformations on the African continent. Although the changes in wildlife numbers may be considered by some to be too recent to influence the evolution of dung beetles, evidence from dung beetles translocated to new continents in the last 50 years reveals rapid changes in their morphology and population structure. Extinction of species is the other, most extreme, evolutionary outcome that could occur, exemplified by the search for the giant *Pachysoma* from Angola. The type specimens of *Pachysoma cancer* in the British Natural History Museum were until recently the only specimens known. This magnificent flightless dung beetle has not been seen for nearly a hundred years, and we have no idea how it has been affected by the depredations of war in the region where landmines may have excluded elephants, which can smell these buried killers. We also do not know with certainty where the elephants' recent ancestors lived or migrated in other regions, or their past numbers, let alone those of the smaller members of Africa's fauna.

Even though the fates of dung beetles and elephants are intricately linked, the fact is that at the level of species different dung beetles go for all sorts of dung, and they do not generally confine themselves to that of elephants. Many field guides and even published research tends to associate certain beetle species with certain dung types. Both the splendid *Kheper nigroaeneus* and the enigmatic flightless *Circellium bacchus* have been labelled as elephant dung specialists, which they aren't. The former is equally adept at using cow dung, and the latter has been found on human faeces, while isotopic evidence suggests it subsists on rat droppings: less romantic, but no doubt more available.

These catholic tastes allow dung beetles to participate in one of the greatest spectacles of wildlife migration still to be seen in Africa, and reveals how busy they can be in moving and working with different species of animals. The annual migration of the wildebeest in a giant loop through the Serengeti/Mara ecosystem (which encompasses an area of 25 000 km², crossing between Tanzania and Kenya) is ancient. Dung beetles play a role in this trek, flying alongside the huge herds of wildebeest and laying down up to three generations of dung beetles in any one migration period. During the wet season, estimates put the volume of wildebeest at 250 per km², which translates into a wonderful wealth of dung for the beetles, especially when you consider that there are approximately 1.3 million migrating wildebeest in total. The dung beetles play a major role in recycling this dung, redistributing nutrients deep into the soil, allowing penetration of both air and rainwater.

The route taken by the wildebeest and dung recycled by the beetles ensures that the area stays fertilised and that new grasses will grow for the rather fussy wildebeest. As with many grazers, wildebeest do not randomly consume what lies immediately under their noses. Instead they carefully select specific grasses in different areas, and return to it to graze when it is at the right height and nutrient content for them. On their great migration north, an armada of dung beetles species work in different areas, differentiating between sandy soils, clays or woodlands. The latter tends to harbour slow-burying tunnelling species, while open grassland favours fast-burying tunnellers and big ball-rollers, which seem to prefer the open sky for both flight and ball-rolling cues. Dung beetles have been estimated to recycle up to 75 per cent of the dung in the Serengeti, with 15–20 per cent of the soil in the system being made up of buried dung balls. There are not many studies on this most ancient route of dung beetles and their friends the wildebeest, which is odd considering it is one of the oldest migratory routes we know of in Africa. This is one place where a sense of the history of dung beetles and their relationship to migrating wild animals could be explored and better understood, including examination of the idea of dung specialisation by certain beetle species. After all, elephant dung (with the appearance of a bolus of wet, fibrous knitting) looks like the last thing any animal, no matter how dextrous, could make into a ball, roll away and eat.

The Serengeti migration also demonstrates something that another 1970s Oxford-trained ecologist spent his life exploring, within different contexts and species (including dung beetles). Ilkka Hanski (1953–2016), was a Finnish scientist who did his PhD at Oxford: this topic was a felicitous product of the first studies undertaken on elephants and dung beetles. Hanski went to Oxford with the intention of following in the footsteps of Charles Elton, the father of invasion biology. Here the head of the zoology department encouraged Hanski to look at the largely unexplored world of mites instead, but Hanski fortuitously met Malcolm Coe shortly after he had returned from his work in Africa. The outcome was that Hanski decided to study local dung beetle populations. His choice of experimental subject was either serendipitous or inspired for someone seeking to understand how biodiversity is maintained. The community of dung beetles (even in Europe's limited fauna) is an excellent one in which to explore population dynamics, a subject that interested Hanski throughout his years of research. Sharing a clearly defined habitat (the dung pat), subsisting on the same food, and being driven by the short lifespan of the dung makes the dung beetle community ideal for modelling how habitat fragmentation influences population extinction events.

Hanski, who is now known as the father of metapopulation theory, was always seeking answers as to how small invertebrates spread and survived loss of habitat. He undertook research on a variety of creatures, most notably a butterfly found on an island off his native Finland. He then conducted research in Madagascar, using its isolation to explore its dung beetle population, and the way that dung beetles had colonised the island and its dung resources. Due to the island's remoteness, the dung beetle community there lacks the more recently evolved species. Madagascan dung beetles have also adapted to the absence of big native ungulates; instead, the largest indigenous mammals are lemurs. Even though opportunities now exist for the indigenous dung beetles to explore new food sources, such as cattle dung, and new habitats (such as pastures) it appears that these dung beetles have not in general been able to exploit new opportunities. They appear instead to be relatively conservative, which may or may not work to their advantage given that rapid habitat loss is probably the greatest constant threat to any organism in Madagascar, including its human dwellers. (This is probably the most significant issue affecting our biodiversity across the globe).

Hanski went on to influence a new generation of researchers in the northern hemisphere; meanwhile dung beetle research was gathering momentum in South Africa. The difference between South Africa and the territories to its north is as much a product of political as historical factors. Lying on a trade route to the lucrative east, South Africa was one of the first territories where extensive collecting of flora and fauna took place in the seventeenth and eighteenth centuries. It is also one of the few countries where traces of dung beetles are not only found in nature, but in some of its cultural treasures. One of the most famous literary appearances of a dung beetle is in Olive Schreiner's classic novel, *The Story of an African Farm*. An early feminist novel with a complex structure, it narrates stark portions of the difficult lives of three people, first as children, and then later as adults. The following excerpt is typical of the tone of the book:

The dog watched his retreat with cynical satisfaction; but his master lay on the ground with his head on his arms in the sand, and the little wheels and chips of wood lay on the ground around him. The dog jumped on to his back and snapped at the black curls, till, finding that no notice was taken, he walked off to play with a black beetle. The beetle was hard at work trying to roll home a great ball of dung it had been collecting all the morning: but Doss broke the ball, and ate the beetle's hind legs, and then bit off its head. And it was all play, and no one could tell what it had lived and worked for. A striving, and a striving, and an ending in nothing.[5]

Schreiner portrays the pointlessness of the beetle's struggle as an existentialist metaphor for the pointlessness of labour and life. This is the polar opposite of Khepri and the hope of transformation, and an ironic metaphor (given that the dung beetles are prime examples of how that apparently pointless struggle to survive is what shapes evolution).

The long history of engagement with wildlife in South Africa, both destructive and constructive, has meant that it has been a place where studying the local flora and fauna is a well-established tradition. Another consequence is that although largely confined, South Africa's wild animal populations are relatively healthy. While a pariah state through the years of apartheid, South African scientists were nevertheless among the leaders in many scientific fields, which might explain why, when the Australian dung beetle project was extended beyond that country, South Africa was one of the places chosen to locate their dung beetle research. The climate, beetle diversity, good infrastructure (with regular flights between the two countries), the presence of an Australian High Commission (and later an Embassy), undoubtedly also all played a role in this choice. When the Australian CSIRO Dung Beetle Research Unit (DBRU) was set up in the 1970s in South Africa, it became the focal point for an ever-expanding number of studies on dung beetles in Southern Africa. It was also the incubator for the career of several dung beetle specialists.

George Bornemissza, the instigator of the Australian dung beetle project, was fittingly the first director of the DBRU, which opened in

Pretoria in 1970. The unit ran under the patronage of the Australian Embassy in Pretoria, which not only administered the researchers' wages but also invited the renowned Dr Bornemissza to embassy functions. Legend has it that at one such meal, the ambassador's wife was concerned to see her distinguished guest removing elements of the salad before eating the remnants. Enquiring if the doctor did not like that particular ingredient, the hostess was shocked to learn that it was the beetles, rather than the good doctor, that might object to the contents of the meal.

Notoriously a little eccentric (as many entomologists are), Bornemissza ran the unit for nine years, during which he supervised the collection of tens of thousands of dung beetles from all over Southern Africa, Europe and China. The unit also exported thousands of surface-sterilised dung beetle eggs to Australia where, in quarantine, they were reared to adulthood. These were then mated and employed as the parents of the next generation of beetles that would eventually leave quarantine (in the form of surface-sterilised eggs) to be reared, released and set to work on the continent's fly problem. The procedure obviously involved unpacking the beetles' eggs from their natal brood balls, because the dung might have carried numerous unwanted stowaways, including foot-and-mouth disease. As a consequence, sterilised beetle eggs had to be repackaged into hand-made brood balls, which must have made the CSIRO insectary look a little like a boutique chocolatier. Eventually, a ball-making machine was built.

More than 50 species were introduced into quarantine in Australia, and 43 species were released, of which 23 became established. Some species are now so widespread that they have common names in their new home, such as the blue Christmas beetle – not a whiff of dung in that one. The first releases in Australia in 1967 were predatory hister beetles;[6] these had been early introductions into Hawai'i, which were then passed on to Australia because Bornemissza had shown *Hister chinensis* to be a maggot killer in Fiji. But the beetle that really sold the project to the Australian cattle ranchers was *Onthophagus gazella*:[7]

a pretty, chestnut-coloured African tunnelling species, about the size of a marble. Introduced in 1968, followed by four more African species arriving via Hawai'i, the tiny-horned males and industrious females proliferated. They packed away enough dung to convince Australians, who had seen biocontrol success with the Cactus Moth controlling the prickly pear cactus (but had had their fingers burned with the Cane Toad eating everything in sight, except cane beetles), that Bornemissza's plan was viable and safe.

Back in Africa, despite the heat and the dung, Bornemissza insisted that his male staff wear a jacket and tie to work when at the unit. He preferred that field equipment be made from surgical-grade stainless steel, and was equally meticulous about building a world-class collection of dung beetle specimens; a collection which then attracted other entomologists who were keen to learn, and also to describe new species that were being swept up into the immense collecting efforts of the unit.

Stories circulated about some of these fanatical taxonomists who devoted themselves to cataloguing[8] the DBRU's enormous dung beetle collection, and became so possessive over 'their beetles' that some of them couldn't resist departing with type specimens in their luggage. Here it is vital to understand that a type specimen is the actual individual of a species to which the name and description of that species is attached. It is therefore the physical representative of that species, and an important piece of scientific information and history. Taxonomic crimes don't come any bigger (except perhaps for the wasp taxonomist who was reputed to cut bits off the types when they didn't agree with his classification system, and apparently threw a whole series of types overboard for the same reason).

A similar crime was perpetrated at the DBRU. Specimens were literally lost by being thrown out of a window by the unit's preparators, who got fed up with pinning what looked like the same species over and over again. This entomological delinquency was uncovered only when a staff member discovered piles of beetle corpses under the window of the prep room. Given that many dung beetle species (and other insects) can only be distinguished by microscopic examination of the male's aedeagus – the insect equivalent of a penis – a few new species might well have been lost along the way.

Bernard Doube replaced Bornemissza at the DBRU in 1979. A native Australian, he had a more relaxed attitude to dress codes. He was also a more modern experimental ecologist who collaborated widely, researching the effect dung beetles had on flies breeding in dung using the African buffalo fly as a model for the pest species in Australia. Fly breeding success was reduced by up to 90 per cent in Africa by dung beetle activity, compared to 66 per cent in Australia. So the DBRU redirected its efforts towards understanding what was actually happening in the dung pats where flies and beetles were competing for a shared resource, while both being preyed upon by other species of beetles. Fieldwork was conducted in Hluhluwe Game Reserve in KwaZulu-Natal, where some of the dedicated researchers lived in a caravan for two years. More permanent accommodation was eventually built, which has since been donated to the park – it is still affectionately known as 'Dung Beetle'. For two years, it housed a full-time research technician who sampled beetles monthly, from not only the Scarabaeidae, but also the other three families of beetles that frequent dung.

Later efforts saw buffalo fly eggs being transported over 500km from Pretoria down to Hluhluwe every week, where they were then carefully transferred onto buffalo dung pats using a fine paintbrush. This was meant to tempt the predatory beetles, and in particular the staphylinids, to eat the fly eggs. Staphylinids are another family of dung-dwelling species, commonly known as rove beetles. The most famous is the Devil's coach-horse beetle, which looks nothing like a horse, and more like an earwig than a beetle. In theory, the fly-egg proteins could be detected in the gut of the predator, using an antibody raised in rabbits, and they could then be labelled as specialised predators of the target fly. Dung beetle research had suddenly leapt into the modern realm; unfortunately, the technique did not work in this case.

As a consequence, the staphylinids never made their way to Australia before the research unit was closed down in 1986. The Australian Meat Research Committee, the principal funder, had decided to invest elsewhere (into meat packaging, as it happened), and the scarab beetles that had been imported into Australia had yet to build up sufficient numbers across that continent to show their true worth.

Jane Wright oversaw the last days of the DBRU in 1986. One of its lasting legacies was the collection of more than 50 000 specimens of 600 dung beetle species, which were eventually donated to the South African National Collection of Insects in Pretoria. The other, more widely felt, legacy are the 23 species of introduced dung beetles successfully established in Australia, from which 1.7 million dung beetles were bred and released into the Australian continent.

If any one particular beetle encountered by the DBRU researchers over the years deserves special mention, it is *Kheper nigroaeneus*. A large ball-rolling beetle occurring over much of Southern Africa, it is a lustrous, deep metallic, purple colour. Easily as popular as its godly namesake with dung beetle researchers, it looks the part and can do it all. It rolls balls, orientates by the sun, makes nuptial gifts, shows extreme maternal care, shifts dung in spade-loads, but never made it to Australia – possibly because the adult female is needed to nurse the larva through to adulthood. All these endearing behaviours were deciphered by careful tracking of individual beetles from year to year; periodically digging up nests that had been flagged in previous seasons to discover exactly what was going on underground. Fabre would have admired the painstaking methods of Penny Edwards and Hartmuth Aschenborn of the DBRU who, after unravelling the life and loves of *Kheper nigroaeneus*, also unlocked the triggers for dormancy hindering the introduction of some other useful dung beetle species destined for Australia.

The DBRU closed quietly in 1986. Adrian Davis, who had been one of the Bornemissza's earliest employees (joining the unit after arriving on an overland trip from Britain) continued his dung beetle career at

Pretoria University. Davis knows more about African dung beetles than anyone else in the field and continues to describe the distribution patterns of dung beetles in terms of biogeography and community structure. Marcus Byrne shook hands with Davis as they locked up the unit for the last time, and Byrne went on to Wits University where he slowly realised what three years of working with dung beetles and dedicated scientists at the DBRU had taught him about science.

The larger Australian dung beetle programme did enjoy a brief renaissance from 1990–1992, when the CSIRO department of agriculture in Western Australia released two European species of dung beetles, introduced as adults via their ultra-high quarantine facility at Geelong. These, along with the African beetles, have undoubtedly improved pasture quality across much of Australia and all but removed the bush fly from high rainfall areas of southern Australia at certain times of the year. Street life has blossomed, to the extent that a café society has emerged in cities like Canberra, where sidewalk food had previously needed to be served behind extensive netting. Unfortunately, there has been a less dramatic impact on the buffalo fly; it breeds in massive numbers early in the new season, before most of the dung beetles have got going. This has stimulated another Australian dung beetle revival, led by Jane Wright in Canberra from 2012–2015; this has resulted in the introduction of *Onthophagus vacca* and *Bubas bison* from Europe, in the hope these cool climate tunnellers will bury the early season flies before their numbers take off. Nevertheless, the importance of dung beetles as architects of soil health has again been recognised, with a new multi-million dollar project announced in 2017 by the Australian government to research dung beetles' contribution to soil conditions.

Across the Tasman Sea, antipodean neighbours in New Zealand released 11 species of dung beetles from Australia in 2011, three of which are of African origin with the rest either from Europe, America or Australia. These tunnelling introductions are expected to improve soil health rather than undertake fly control, and will augment the flightless, indigenous New Zealand dung beetle species, which are restricted to shaded habitats and avoid cattle droppings.

The DBRU in Pretoria was a fun place to work; practical jokes come naturally where dung is sufficiently abundant to bury someone's car keys in 40 litres of it. The Australian programme, along with its South African field station, straddled another period in changing fortunes of dung beetles. The initial goal of the programme was pure economic entomology. Tens of thousands of beetles were collected, bred and released for the benefit of agriculture; a few scientific papers were published, and no one complained. But this research into the ecology and behaviour of the beetles also revealed their other talents: in sexual selection, in flexible hereditary and in complex community structures (all sensitive to environmental change). Other teams of beetle researchers in South America, the US and Europe were now staring hard at their own dung beetle fauna and realising just how many important ecological and behavioural questions could be answered by the descendants of Khepri. The dung beetle world was rapidly expanding in a flurry of research papers, multiplying annual output tenfold post-1986 compared to the preceding 26 years' output.[9] By comparison, research papers on drosophila, the tiny vinegar fly about which we know more than any other organism on earth, increased only five-fold (half that of the beetle increase) in the post-1986 years. After a three-thousand-year hiatus, dung beetles were back in fashion.

CHAPTER SIX
Tribes with human attributes

THE INCREASING SCIENTIFIC INTEREST in dung beetles was fuelled by both the burgeoning scientific industry (with biology diversifying into a multitude of new fields, from invasion biology to genetic engineering) and by the usefulness of dung beetles in our explorations of these new fields. Two characteristics of dung beetles have made them stars of the new biology: first, there are lots of them – over 6 000 species, offering a huge variety of already named species on which to test new ideas; secondly, they are incredibly compliant. Dung beetles, whether in captivity or the wild, get on with their lives regardless of who is watching or manipulating their small world. This makes them perfect subjects for studies of animal behaviour, in which we can ask questions of a simple animal and get straightforward answers, which in turn inform our understanding of how other organisms (including ourselves) perceive the world.

Why do animals move? For what purpose? The world is full of animals with which we share common spaces. Most of those animals move, apparently with reason or purpose, leading us to wonder what they are up to. These questions are at the core of the study of animal behaviour which, despite being an ancient human activity, is considered to be a relatively young science. Nevertheless, being formalised as a science under the rubric of 'ethology' by French biologist Isidore Geoffroy Saint-Hilaire (1805–1861) gave credibility to the careful observation of nature often provided by enthusiastic collectors. Darwin treated behaviour (along with structure) as a characteristic subject to natural selection, which could therefore influence an organism's chance of survival and reproduction. Unfortunately, Saint-Hilaire chose a word deriving from 'ethos' (from the Greek for 'moral character') and with a history of other meanings. Seventeenth century actors who portrayed human characters on stage were known as ethologists, while John Stuart Mill used the word ethologists to describe those who studied ethics. The name ethology has nevertheless stuck, and is used to describe the study of the behaviour of animals in their natural habitat.

Fabre (despite his rejection of a Darwinian interpretation of why the natural world looked the way it did) made important contributions to the infant science of ethology through his detailed observations of the behaviour of dung beetles and other insects, including bees and wasps. The first obviously ethological publications were studies of bird behaviour, published in 1911 and 1919. Because birds are easily seen and have stereotypical behaviours, it is not surprising that they were the main subject of study for Konrad Lorenz and Nikolaas Tinbergen, who worked in Europe in the 1920s and 30s. Nevertheless, the man with whom they shared the Nobel Prize in 1973 worked on bees. Karl von Frisch (1886–1982) devoted his life to the study of their behaviour 'thus discovering a true language of gestures for communication and opening new insights into the knowledge of insect behaviour.'[1] It is worth noting that this joint Nobel Prize was awarded for Physiology and Medicine (there is no Nobel Prize for Biology) because an animal's behaviour inevitably develops within the bounds of its physiology and morphology.

The close alliance of ethology with morphology helps explain the emergence of this new approach to science in the nineteenth century, following on from meticulous descriptions of both animal and plant anatomy. Comprehensive descriptions of animal behaviour (such as those by Fabre) began in the nineteenth century, to be followed in the twentieth with attempts to understand the mechanisms underlying that behaviour. Dung beetles and their study have made major and exciting contributions in both areas.

Why dung beetles? As explained above, one reason is simply because there are so many of them. We have given names to at least 6 000 species worldwide, and they have the potential to converge in massive numbers on one dropping, with up to a hundred species competing for exactly the same resource simultaneously. As a consequence, dung beetles are the ideal organism to study where comparisons between species will reveal how and why a particular species has evolved to overcome a particular environmental challenge. Asking how species A uses dung compared to species B will reveal what mechanisms and behaviours are important for that task. The fact that the true dung beetles are all closely related, being in the same subfamily (the Scarabaeinae), makes that comparison even more revealing because they have all inherited similar beneficial adaptations and deleterious deficiencies from their common ancestor. In addition, all the dung beetles in the same dung pile are eating the same dung, in (apparently) the same way, in precisely the same habitat. It gets better. Dung beetles carry out their daily chores oblivious to any would-be ethologists with their notebooks, cameras and cunning experiments designed to unpack the causes and function of their singular behaviour – wadding dung into balls and stashing it underground. Bees and ants, the favourites of the insect ethologists, don't come close to the dogged tenacity of the dung beetle, which doesn't need training or motivation to give answers to questions such as: what are you doing? Can you see colours? What are you looking at? How do you roll a ball of dung in a straight line? Why make dung into a ball to roll across the surface of the earth?

Obviously the first question is: why do dung beetles roll balls of dung? The easiest answer is that they have evolved this behaviour because somehow it enhances either their chances of survival or successful reproduction. As with many questions about animal behaviour, a side issue emerges – why should we care? Well, given that dung beetles represent many small and largely unnoticed organisms at the base of just about every food web, understanding the needs of the earth's recyclers is critical to ensuring that they continue to liberate essential minerals and nutrients from the detritus (dead stuff) that would otherwise overwhelm the living and thereby halt their growth. Besides, it is intriguing. Finding out how other animals make a living – out of dung of all things – is fascinating.

Human ingenuity can take discoveries from nature and almost without fail turn them to our benefit. The structure of the atom (nuclear power) and cockleburs (Velcro) are two obvious examples, but sharkskin (drag reduction), butterfly wings (interference colours) and fireflies (light amplification) are more recent (and entomological) examples of this kind of discovery. Dung beetles, as we have seen in the previous chapter, offer us the opportunity to manipulate ecosystems – in the biological control of dung-breeding flies and nutrient re-cycling. However, their peculiar ball-rolling behaviour (accomplished with precision using a tiny brain) also points to advances in robotics, where seemingly complex challenges such as orientation and navigation can be accomplished with minimal computing power if insect brains (like the dung beetle's) are used as models. Paring their task down to its bare bones, a dung beetle can roll a ball in a straight line using an in-built celestial compass with fewer nerve cells than you have in your fingertips.

But to return to the first question, why do dung beetles roll balls? Most of them don't, in fact, opting rather to bury the dung directly beneath the pat. So why do ten per cent risk predation and theft by exposing themselves on the soil surface, and adding a directional task to their job? Competition seems to be the most obvious and satisfactory answer.

Space beneath a dung pat ends up being severely limited because up to 90 species of dung beetles can be found tunnelling and storing dung for feeding or breeding under a single pat. Defending the dark recesses of one's dung stash under such circumstances is difficult, especially when one is unable to patrol the area around the underground perimeters of the buried dung. Rolling the ball away avoids this bunfight, but opens up another area of competition from fellow rollers who might want to cheat by stealing a ball rather than invest time in making their own. Under these conditions, the competition is evident to any observer. Beetles flip each other off balls like bouncers wading into a bar-room brawl. With a flick of their powerful front legs, the owner (usually) repels all boarders, sending them cartwheeling off into the grass – unless the aggressor is hotter (literally). By measuring the body temperature of winners and losers in these battles, researchers have shown that the warmer bodies generally win the fights.

Most insects are ectothermic; their body heat comes from an external source, which means that the environmental temperature usually dictates their body temperature. But some species are able to raise their body temperatures metabolically to fly, or to fight. A dung beetle flying onto a busy dung pat has the option to keep its temperature (elevated for flight) at the raised level, and therefore stands a better chance of being successful in a fight over another beetle's ball. If the new arrival allows itself to cool down, then the better energetic strategy would be to make a ball, with the hope that it is not stolen by a hot thief.[2]

Being hot can have its limitations though, even for a ball-rolling dung beetle. The soil surface temperature of an African savannah can exceed 60°C. Given that most animals (ectothermic or otherwise) go into some sort of heat shock at around 40°C, even a large ball-rolling dung beetle weighing under five grams cannot spend long on the hot sand before its body temperature starts to climb towards lethal levels. Evolution has addressed this problem by co-opting an existing behaviour, the orientation dance, into another behaviour – thermoregulation. Researchers working on dung beetle orientation noted that on hot days, the dung beetles rolling their balls across the baking sand danced more frequently. By occasionally climbing on top of their balls during normal rolling behaviour, the dung beetles scan the sky for orientation cues during a 360° rotation behaviour dubbed the orientation dance. On hot days, however, the dance involves more scratching and head-wiping and apparent irritation than usual. By using a thermal camera, which sees heat rather than light, the scientists could then observe that as the ground got hotter, so did the beetles; consequently, they spent more time perched on top their dung balls to cool off. The moist dung remains colder than the ground through evaporative cooling, offering a thermal refuge to an overheating beetle, which will periodically climb down to plunge its head and front legs into the baking hot sand to lever the ball along a few more centimetres. By putting little silicon oven gloves on their subjects' front feet, the researchers were able to reduce the frequency of heat-induced dancing by the beetles, showing that this behaviour was mediated via sensors in the front legs. Climbing off hot soil onto some sort of thermal refuge is called stilting, and is seen in other insects (such as ants), but dung beetles are the first animals other than humans known to bring their own thermal refuges along with them:[3] (we wear sandals on the beach, or spare our roasting feet by standing on a towel before rushing across the hot sand into the cool sea).

The beetle dance is primarily associated with their orientation behaviour, and might be used to set a straight-line course while rolling a ball. However, evolution has repurposed this dance as an additional function to help regulate body temperature; in the same way the dung ball has been co-opted, not only as a thermal refuge, but also as a nuptial gift when courting a mate. Chocolates or flowers work for some people. Dead insects wrapped in silk are de rigueur when wooing female scorpion flies or dance flies. But, predictably perhaps, some male dung beetles use dung balls as a reward for sexual favours from females they attract to a tryst with pheromones.[4]

Given that smell is clearly how dung beetles initially locate their food, and knowing that sniffing out a potential partner is how dung beetles get to meet their mates, the olfactory world of dung beetles is unexplored by science to a surprising degree. Adult dung beetles are highly sensitive to smells, as evidenced by the relatively huge antennal lobes in their brains which are associated with olfaction. They also have several sets of pheromone glands distributed over their body, some of which have been studied.[5] The majority of these glands, although described, are of unknown function. Communication via olfaction is apparently a big part of the dung beetles' world, and wide open to further research.

Pheromone attraction of females by head-standing males happens either publicly at the dung pat or more intimately at a burrow entrance, depending on the availability of females (which become increasingly scarce as the season moves on). This is because many of the females in dung beetle genera that exhibit maternal care (such as *Kheper*) will be back underground, tending their brood balls to ensure the survival and development of the larva within them. Solitary males still found at dung pats on the surface are now competing for fewer females rather than for the dung, which is still being removed by rival species.

The *Kheper* males call to sexually mature but unfertilised females by means of a pheromone, a chemical message that leaves the beetle's body as a white waxy powder. The message is clear: join me for sex and I will reward you with a dung ball and sperm. The calling male uses a tuft of stiff little bristles on his back legs to brush the waxy pheromone off the side of his body, which allows the pheromone to drift away in wisps into the air. All the while he poses head down in the dung or his little tunnel, hoping that a passing female will respond to the leg-jerking and freezing postures that accompany his performance. A small ball of dung (large enough to eat, but not to lay an egg in) is the usual reward for the female lured to such a male, who may or may not take any further part in the process of rearing the offspring once he has inseminated her. Given that they can get away with offering such cheap dates, some males push their luck and even cheat: they call from an empty burrow, duping the female into free sex, with no dung ball to offer. This might be one of the many dung beetle behaviours we humans recognise and find parallels with in our own lives, but it is hardly endearing in the way their industry and tenacity is.

Finding a suitable mate is tough in any corner of the animal kingdom, and dung beetles obviously have to make compromises like the rest of us. Male dung beetles (as with most males of most species) will generally benefit in evolutionary terms if they can inseminate multiple partners, thereby increasing the chances that their genes will be passed on to the next generation. Because this conflicts with the behaviour of the females, which invest many more resources in maternal care of fewer individual offspring, sexual selection of male dung beetles by females is often strong. If her mate won't help raise the children, then he had better bring some good genes to the party.

Darwin recognised the evolutionary power of sexual selection (the chances of getting chosen as a mate) in addition to the processes of natural selection (the probability of surviving long enough to mate). Some male ball-rolling beetles, such as the flightless *Circellium bacchus*, basically use their bulk to win access to females at the dung pat. Smaller males, through no fault of their own, lose these fights primarily because they had a limited amount of food in their natal brood ball (size having a low heritability in dung beetles). Their parents (usually the mother) simply couldn't make them a large brood ball, probably because dung was in short supply at that time. This, unfortunately for their offspring, would have restricted the food available for that individual's growth and development, a shortfall that would dictate how it would develop inside the brood ball and in life beyond. In the case of non-rollers, the developing male larvae, sensing the amount of food available to them, will either compromise on basic body form (going without ornamentation such as horns), or reflect their parents' superior provisioning skills by emerging from a large brood ball as a major male, adorned with horns that can be half their body length in some species.

Most ball-rolling dung beetle species don't have horns, probably because they don't have to guard females in tunnels from their neighbours in the crowded metropolis under a dung pat. Instead, having rolled a potential brood ball away to a more secluded spot, ball roller males can monopolise a female without fear of being cuckolded. Major and minor males of ball-rolling species differ primarily in size, and conduct their displays of machismo on the surface (at the dung pat itself) where smaller males will generally lose a fight. However, rather than investing limited larval resources in bulk, *Circellium bacchus* larvae growing inside a small brood ball develop relatively larger testes than their beefy brothers. Then, if a smaller male does manage to sneak access to a female as she buries the brood ball, and her buff consort is not concentrating, he increases his chance of achieving paternity by transferring a larger amount of sperm into the female.[6] She has little to lose by cuckolding her partner because the adulterous minor male is just as likely to have good genes as his larger rival. In those species where males carry horns, large females generally do not have them. Instead, bigger females get a boost in the natural selection stakes by being able to produce more and/or larger eggs, getting a greater number of healthy offspring into the next generation that way.

Tunnelling dung beetles are often sexually dimorphic, as described above. The major males of a particular species usually carry horns as secondary sexual characters that are not directly involved in the process of reproduction. Some examples are (in relative terms) as impressive as the majestic curling horns of a mature male kudu. As in the kudu, the beetle horns are used in fights between males over females, or territories where females can be found or monopolised by the males. Only large males from big brood balls have the resources to invest in such expensive headgear, sometimes even at the expense of having relatively smaller eyes than the minor males[7] – confirming that even for dung beetles, there's no such thing as a free lunch. Smaller minor males in some species once again resort to sneaky copulation, in this case by digging a side tunnel around the major male who sits blocking the front door access to the female with his huge horns. The back-door approach to the female may be to her advantage too, increasing the variety of sperm she receives and the genes she is able to pass on to her offspring. However, as the number of sneaks in her locality increase, the size and number of brood balls produced by a female decrease, suggesting that she is so pestered by sneaky males that she cannot concentrate fully on her maternal tasks.

As with many other organisms, both animals and plants, female dung beetles invest more than males when it comes to getting their offspring to adulthood. Females of smaller species, such as *Euoniticellus intermedius* (one centimetre long), usually make a nest with up to ten small brood balls, each about the size of a grape. Into the hollow centre of each ball, the female lays a single egg. She achieves this neat trick by first making a cup at the end of a tunnel branch, using dung collected from the pat at the soil surface above. The paternal male might assist her by dragging dung into the main nest tunnel, although this behaviour might be curtailed if a large number of other males are lurking in the same dung pat: paternal care in dung beetles ranges from zero to full brood care, depending on the species and the immediate neighbourhood. Once the cup is formed, the female *Euoniticellus intermedius* then backs into it, carefully depositing her relatively huge egg onto a specifically prepared spot in the bottom called the maternal gift. This is a smooth patch (or pillar in some species) made of an unknown substance that supports the egg, holding it away from the wet inner wall of the ball that might otherwise drown the developing egg. The female beetle then climbs out of the cup and skilfully folds in the upper edges to give the finished brood ball the structure of a Fabergé egg, with a concealed gift inside it.

The composition of the brood ball is remarkably similar to the raw dung from which it is made. Although adult dung beetles are very selective when eating dung, choosing only the smallest dung particles, the brooding female uses the full range of particles available in the dung to construct a brood ball. Nevertheless, she makes one potentially important change to the nutrient status of the brood-ball dung: her maternal gift has an additional role as a macerated first meal for the newly hatched larva. So in addition to holding the egg in place, the maternal gift has a higher nitrogen content than the surrounding brood ball. Nitrogen is an essential component of protein, which the developing larva will use to build its young body. Although newly hatched dung beetle larvae have sharp, toothed jaws (which the parents lack), the head and the associated jaw muscles are small. The tiny particle size and elevated nitrogen content of the maternal gift are presumably consumed without much effort: the beetle equivalent of the smooth puree with which we introduce babies to solid food. Despite the importance of this maternal gift, we still don't know from which end of the female it is produced, or what she adds to its contents.[8]

Once the larva grows in strength and size by moulting, it proceeds to chew its way through most of the brood ball dung provided by its parents. This is achieved using strong jaws reinforced at their inner edge with chitin, a substance similar to the keratin in our hair and fingernails. The jaws are renewed twice (each time the larva moults into the next larval stage), but then are lost at the final moult, being completely remodelled into soft hairy mouthparts in the adult beetle – incapable of chewing the larger plant particles that make up the bulk of herbivore dung.

The new adult beetle emerges from a pupal cell, which is the remainder of the brood ball, reinforced by faeces ejected from the last-stage larva as it empties its gut in preparation for pupation and metamorphosis: a radical reorganisation of its body as it transforms from something like a worm into a hard shiny beetle. This transformation belies the intriguing fact that despite both eating dung, adult and larval dung beetles have completely different diets. The larvae chew the larger, harder components of the brood ball dung, while the adults consume the smaller, softer components of fresh wet dung. This represents a division of resources almost as extreme as that of a caterpillar chewing plant leaves in contrast to the adult butterfly it will become (which will subsist by sipping nectar).

The consequences for both adult and larval dung beetles of dividing the dung cake between them, are enormous and are only just beginning to be explored by scientists. The adults effectively sieve the dung to concentrate the nitrogen-rich small particles, used for building proteins such as muscle. The dung selected this way and swallowed by the adult beetle, when extracted from the foregut by researchers, has been found to be up to five times richer in nitrogen than the original dung from which the beetle has fed. This selective feeding transforms a useless food source such as dung (containing about one per cent nitrogen) into an acceptable meal of about five per cent nitrogen which would be more than sufficient to sustain the average herbivore, either insect or mammal.

How then does the larva cope? When the maternal gift is gone, the larva only has big bits of fairly dry, low-nitrogen dung to contend with. Worse, the larva can't feed selectively like an adult because the brood ball is all the dung it has until it metamorphoses into an adult and can find its own food. How will the brood ball meet the high demands of a developing body that needs nitrogen for protein as much, if not more, than it needs carbon for energy? As yet, we don't know the answer. We do know that the larva chews, digests, defecates and re-eats the brood ball contents over and over as it rotates inside its pupal dwelling, which has been likened to living inside its own rumen – the fermentation chamber in the guts of bovid antelopes in which plant materials are broken down by microbes. It remains to be discovered how valid this analogy is.

Analysis of the assortment of micro-organisms living inside the gut of the larval dung beetle show that they are somewhat different to those varieties found in the parents' guts. This is not surprising, given that they eat different foodstuffs. Whether those micro-organisms contribute to larval digestion and nutrition remains to be seen. The larval microbiota are similar to that of the raw dung, suggesting incorporation of organisms from the dung into the larval gut, and we know ruminants use symbiotic bacteria for digestion of plant material. But the larval beetle microbiota are also similar to the maternal beetle biota (less so the paternal mixture), suggesting that the maternal gift could be a source of useful micro-organisms. What these micro-organisms do, either in the gut of the larva or on the inside wall of the brood ball, we don't know. What will be exciting to discover is whether they boost nitrogen (by fixing it from the atmosphere) inside the carbon-rich nitrogen-poor globe of the brood ball, helping the larva satisfy its needs for protein as it grows a new body.

Our exploration of the many fates of the dung ball – food, brood or gift – explain to some extent why dung beetles package dung into balls and make them round. What still needs to be described is how they carry out this entertaining behaviour. The tunnellers do all this work in private, concealed in subterranean tunnels beneath the dung pat. But the inventive researcher can peek in at the process by confining a mated female *Euoniticellus intermedius* in a soil sandwich, held between two sheets of glass a centimetre apart to create an underground aquarium. Undeterred, she will dig a brood tunnel, bulldozing soil to the surface with her head and front legs until she has a small chamber at the end of a ten-centimetre sloping tunnel. She then drags down gobs of dung from the pat, returning underground facing backwards or forwards, to pack dung into the end of the tunnel in the form of a cup that will be turned into a brood ball. Gravity is probably the main stimulus guiding her subterranean decisions on where to shift the dung and make a nest.

Meanwhile, above the soil, the ball-rolling species have a more challenging task. To avoid other thieving beetles at the dung pat, rollers need to escape efficiently with their prize to be certain that they won't accidentally return to the fracas and repeat the struggle all over again. An efficient escape involves travelling in a straight line away from the dung pat, for which a form of compass is needed.

Just about every organism needs to move at some time in its life. To do this in a directed fashion, it requires an external reference point: a compass cue it can use to monitor its direction of travel. Landmarks like trees and buildings will work as external references to some extent, but they can disappear or change over time. More disturbingly, they appear to move as one moves relative to them. This parallax shift between the observer and its surrounding objects can introduce orientation errors, resulting in curving or even circular escape paths. This, along with the need to remember the nature and relative position of landmarks, suggests that dung beetles (with their average-sized insect brains of less than a million neurons) do not orientate using landmarks. Humans can recall the position of their homes in relation to that of shops, schools and other places important to their survival and reproduction. So can wasps and ants, which use landmarks for journeys to and from their nests, so it is unlikely to be lack of mental capacity alone that prevents dung beetles from orientating by landmarks. However, their "mushroom bodies", structures known to be the seat of memory in insect brains, are relatively small. In contrast, bees and ants have massive mushroom bodies dominating their brain volume in the same way our frontal lobes fill the bulk of our skulls.

Scientists have tested the role of landmarks in dung-beetle orientation, finding that they ignore them, including the dung pile from which they escape with their prized dung ball. This is because when dung beetles fly to a dung pat from a distance of ten kilometres or more away (to a place they have never been before, and will never see again), learning the relative placement of landmarks will strain an already limited neural system. That dung pile will never be revisited by that beetle, and by the time it emerges from underground after either feeding on the ball or brood-caring its larva, the dung pat will be long gone. Learning the layout of its surroundings is just not worth it for the average dung beetle.

Instead of landmarks, dung beetles use celestial cues in the sky to set their escape route along a straight path. Cues in the sky support a reliable compass for many reasons. When these cues change position, they alter slowly and predictably as the earth spins through the galaxy. These cues are far enough away not to suffer from parallax error; and there are plenty of them.

But how does an eager experimenter ask a dung beetle if it is looking at something in the sky? The solution is easier than we might think. We have already seen how beetles will continue their rolling behaviour with their feet encased in silicon mitts. Why not give the beetle a golf cap to prevent it looking at the sky, then watch what happens? Popping a peaked cap on to a dung beetle's head isn't as easy as it sounds – *nothing* sticks to a dung beetle. Nevertheless, once in position, the cap doesn't interfere with the beetle's eyes or rolling behaviour, but once deprived of a view of the sky, the poor creature goes around in circles until the cap is removed. So seeing the sky is required to keep the beetle on a straight path with its ball.

What the beetle is looking at in the sky can be explored by modifying the sky it can see – for instance, by moving the apparent position of the sun by 180° with a mirror. This causes the beetle smartly to reverse its rolling direction until the mirror is removed, at which it returns to its original bearing.[9] Other sky modifications are more difficult to create, but with a bit of ingenuity, the normal light intensity gradient can be reversed with a filter. The natural sky is brighter where the sun is, darkening (almost imperceptibly for us) towards the opposite side of the heavens. Allowing beetles to roll under a sky modified to be darker in the solar half, opposite to its normal appearance, causes the obedient beetle to perform a 180° about-turn, similar to that caused by the mirror.

Such experiments suggest that ball-rolling dung beetles have a series of celestial cues to fall back on should one be absent – such as the sun being hidden by a cloud. This allows a rolling beetle to maintain a straight-line exit from the dung pile even as the sky's contents fluctuate between cloud and full sun, or if the dung pat is under the shade of a tree. The sun is clearly at the top of this hierarchy of cues, and gives rise to many of the other cues. Consequently, the sun position has to be measured or controlled if other cues (such as polarised light) are to be tested for their contribution to the dung-beetle compass.

Polarised sky light is caused by sunlight being scattered in the atmosphere. This causes an alignment of the angle at which the light waves are propagating. With the sun at its zenith (the top of its journey across the sky), the plane of light polarisation is in concentric circles centred on the sun. Therefore, when the sun has sunk to the western horizon, only one edge of these concentric circles is visible (to the beetles, not to us), running across the arc of the sky from north to south. At dusk, therefore, the plane of polarisation of the sky is linear, aligned across the heavens from north to south. By knowing that this simpler pattern occurs at sunset, researchers can manipulate the polarisation cue using a linear polarisation filter. Such a filter allows only light aligned in one plane of polarisation to pass through it. Aligning the filter pattern with the polarisation pattern in the sky generates no change in response of a ball-rolling beetle, which rolls straight under the filter and out the other side, completely unperturbed by the science going on above its head. However, turning the filter by 90° to the sky will make the beetle (the dusk-active *Scarabaeus zambesianus* in this case) take a right-angled turn under the filter, which it then corrects as it emerges into the light outside of the filter. The single-minded tenacity of a beetle with its ball allows this type of experimentation with dung beetles. Their determination to get on with the job of rolling a ball while scientists rush around with filters and mirrors underlines the utility of dung beetles as experimental subjects; no other animal has been shown to be able to orientate by polarised lunar sky light. Other animals surely can, but even those that have been highly trained would probably be terrified or horrified at such a disturbance of their world, and refuse to co-operate.

The moon as a mere reflector of sunlight is a million times dimmer than the sun. Nevertheless (as with the sun) when it is less than 18° below the horizon it still contributes light to the sky, and this dim light is polarised because of the effect of the atmosphere, regardless of where the light emanates from. *Scarabaeus zambesianus* is more than capable of using this paltry light to cut straight lines across the savannah with its ball of dung.[10] In addition to detecting polarised light, the structure of the eyes of these crepuscular and nocturnal dung beetles gives them the ability to operate in dim light.

As a group dung beetles have intriguing eyes, not least because they are split across the middle into two halves, giving in some species a completely separate dorsal and ventral eye on each side of the head. The dorsal eye, looking upwards, seems to be primarily involved in spotting celestial cues for orientation. This has been explored by temporarily painting out the dorsal eye with liquid paper (Tippex). Beetles blinded in this way behave in a similar fashion to their cap-wearing compatriots: they roll around in circles. Once this covering is scratched off the fingernail-tough eye surface, restoring full vision to the subject, it then straightens and rolls off into the distance. The role of the ventral eye is less clear. Given that the beetle's head is tipped back in flight, this suggests that the ventral eyes, when pointed forward by this head posture, are used to avoid obstacles. They may also function in finding the ball (should the beetle lose contact with it) and in nest provisioning, especially in species in the genus *Pachysoma* that can navigate – moving back and forth between two known fixed points.

Diurnal ball rollers have small eyes relative to their body size, receiving more than enough sunlight during the day to discern at least one orientation cue for straight-line rolling. Nocturnal species, on the other hand, have much larger eyes, usually with a smooth glassy corneal surface that hides the typical honeycomb facets seen on the eyes of many insect compound eyes. The function of the smooth cornea is unknown. Inside the nocturnal eye, the structural theme is capture of precious photons. In dim light, nature's response is always the same: in any animal, from a beetle to an owl or giant squid, the size of the photon capture devices is increased. Bigger eyes in dung beetles contain bigger photoreceptors (called rhabdoms) that make up the dung beetle's retina. In nocturnal species, each rhabdom is about twice the size of that in its diurnal counterpart, and is doubled in effective size again by having a mirror placed at its basal end. This is called a tapetum, and it effectively doubles the length of the tubular rhabdom so that a photon that misses being caught on the way into the eye is then reflected back through the rhabdom; there it stands a second chance of being captured and detected as light. The tapetum causes a demonic red eye-glow in nocturnal beetles if their head is viewed along the beam of a headlamp directed into their eyes at night.

This low-light visual capability is taken one step further by the fully nocturnal *Scarabaeus satyrus*. Larger than its crepuscular cousin *Scarabaeus zambesianus*, *Scarabaeus satyrus* can orientate by the light of the stars alone. The Milky Way (which is the central plane of our galaxy) is particularly obvious in the summer night sky of the southern hemisphere, when most dung beetles are active in the summer rainfall regions of Africa. By collecting dung at night, *Scarabaeus satyrus* avoids competition from its large cousins in the subgenera *Kheper* and *Pachylomera*. There are also plenty of large, nocturnal tunnelling species competing for dung dropped at night, but ball rollers such as *S. satyrus* seem to maintain an upper hand in the contest, as long as they can see where they are going.

Because *Scarabaeus satyrus* has such good dim-light vision, researchers investigating its starlight orientation were meticulous in ensuring that no other visual cues (such as tiny LEDS on video cameras, or even the silhouette of equipment against the faint glow of the night sky) could be used by the beetle for orientation. Instead of observing the beetles going through their paces directly, the scientists instead timed the beetles leaving a two-metre arena surrounded by a high wall that allowed the beetles only a view of the night sky, which included the Milky Way. This indirect measure of orientation revealed that when the sky was overcast or obscured by a beetle cap, the beetles got lost and spent minutes wandering aimlessly around the enclosure, while the researchers could only listen to the tap, tap, tap of tiny feet on the wooden floor of the arena. Sometimes, even getting a breakthrough finding can be very boring when the control treatment is a hopelessly lost beetle wearing a cap. In contrast, beetles that could see the stars left the arena in seconds, accompanied by cheers from the researchers who were betting on when and where their beetles would emerge.

Every budding animal behaviourist fantasises about doing experiments in a giant aquarium or a planetarium. Predictably, the ever-accommodating dung beetles didn't bat an eyelid (they don't have any) at being asked to repeat their timed performances under the Milky Way in a planetarium. They responded exactly as they had done in the field, getting gradually more lost only as different elements in the artificial night sky were removed by switching off their projection.[11] A beetle's eye cannot actually see a star. Although very sensitive to low light, beetle eyes lack the acuity to resolve the tiny pinpricks of a star's light into an image that can be distinguished from its closest neighbour. Without dispelling the romance of a dung beetle using the Milky Way to find its way, further experiments have shown that the beetles actually see a light intensity gradient – like the daytime sky, some parts of the night sky are brighter than others – rather than perceiving individual stars. Testing the beetles in a more basic planetarium (consisting of two arcs of LEDs inside a tent) has revealed that dung beetles don't respond to the patterns of the stars, but rather use the bright patch of the southern end of the Milky Way (or compare the differing the brightness of two halves of the sky) for orientation purposes.

It is interesting that our noble beetle can be fooled by a few LEDs into behaving as if it were under the stars of an African summer night. Even more disappointing, a green LED can act as a direct substitute for the sun (green light is the major component of sunlight striking the earth, and the beetles have many green sensitive photoreceptors in their eyes).[12] Nevertheless, these substitutes allow scientists to test exactly what dung beetles need in the sky to roll a straight path through the bush. And given the hierarchy of known cues, these elements can be swopped or paired to examine how the beetle compass latches onto what cues and when, in order to make a successful straight-line escape with its dung ball.

The answer that emerges from these manipulations is far from romantic – the beetles don't care what the sky looks like. The arrangement of its elements can be completely unnatural, even impossible in the physics of the atmosphere. When the plane of polarisation is turned perpendicular to an ersatz green 'sun' (something impossible in the real sky), a diurnal dung beetle can still use either or both elements to move in a straight line. Probably most fascinating of all (discovered by switching different elements of this Barnum-and-Bailey beetle-circus world on or off at crucial moments in the rolling process), researchers discovered that the dance – the little rotation on top of the ball – is the critical moment when the beetle takes a mental snapshot of the sky it glimpses (real or otherwise). Climbing down to push the ball along, the beetle then tries to match what it sees with the representation of the sky stored in its minimal brain during the dance.

That tiny brain is also yielding to modern research techniques, where individual cells can be pierced with an electrode that records the response of (for example) a single photoreceptor in the eye to a stimulus such as polarised light. Once the behaviour of that cell has been characterised, a dye is poured into the same electrode to stain the cell into which it is inserted. This allows the researcher to identify the exact cell being tested, and then trace it to its origin and destination within the beetles' nervous system. These details are added to a map of the neurons, building what we know of the dung beetle brain.[13] With the physical structure of the brain mapped out, physiologists and roboticists can understand and mimic how so few neural elements can solve the complex orientation and navigation tasks performed by dung beetles.

Navigation is fundamentally different to orientation. It requires more information to accomplish a navigational feat known as dead-reckoning, or path integration. Imagine a beetle that travels back and forth between a dung pile and a nest, carrying fresh provisions on each homeward journey. This is a very different behaviour from the one-way, one-time, hit-and-run style of the typical wet-dung ball roller. Unusual circumstances have probably driven the evolution of this unusual behaviour in a group of beetles from the arid south-western region of Southern Africa. Here 13 species of the genus *Pachysoma* wander across the sand of the semi-arid region, stretching up the west coast of Southern Africa. These beetles are different in many ways. They can't fly – a major handicap for animals that live where the food supply is patchy and ephemeral. They don't make balls, but instead drag dry food (usually pellets of dung) forwards to a tunnel they have prepared in advance. This behaviour requires knowledge of where the food is and where the nest is relative to the food, so that the industrious beetle can scuttle between the two sites, fetching and stashing dung.

Displacing a *Pachysoma striatum* beetle at its forage site causes a characteristic form of navigational confusion. Placing the feeding station on a piece of sandpaper allows it to be slid sideways, perpendicular relative to the nest-food axis, with both the beetle and the dung pellets on it. Oblivious to the disturbance, the displaced beetle heads for home, walking off the sandpaper, dragging a dung pellet in roughly the right direction, but parallel to the correct path between the food and the nest. Having walked the correct distance, the beetle will start circling in search of its nest entrance, which it will never find. The beetle is hopelessly lost because it uses dead-reckoning and not landmarks to navigate through the sandy landscape of Namaqualand. Dead reckoning requires the navigator to measure not only the direction of travel, but also the distance covered on the course. The celestial cue toolkit, shared with all its ball-rolling cousins, is used by *Pachysoma* to plot its outward direction from the nest, which it then reverses on the homeward leg. Distance measurements are another issue, however. We don't know at present how the beetles measure distance walked across the sand. Some clues come from desert ants that count their steps. Researchers showed this by either adding tiny stilts to the legs of their ants or, less ingeniously, shortening their legs by clipping off the bottom section. The ants on stilts overshot the nest, taking the right number of longer-than-normal strides towards the nest. The ants on stumps came up short, putting in the right number of somewhat curtailed steps to finish on the feeder-side of the nest.

Repeating these experiments on dung beetles has all sorts of constraints, including the sensitivities of the scientists unwilling to cut bits of legs off their beloved beetles. Instead, inserting a slippery surface between the nest and food results in a homeward overshoot if the extra steps are accumulated on the outward trip. Slipping on the way home brings the beetle up short of the nest, having counted the slipping steps without making any forward progress. These experiments are still not yet conclusive, but the idea that step-counting is used in dung beetles gets support from another species of *Pachysoma*, *Pachysoma hippocrates*. Cruising across the Namaqualand sand in search of food, this beetle usually runs with a tripod gait, which is typical of the five million or so other insect species that share our planet. Two legs on one side, and one leg on the other are swung forward and planted firmly on the ground, before the opposite set of the middle leg and the front and back leg of the other side are lifted and swung forward together. However, *Pachysoma hippocrates* has an additional gait in its locomotory repertoire. Every now and then, it switches from tripod running to a gallop. Now the legs work in pairs. The powerful front legs (with their broad-bladed tips) pull into the sand like a butterfly-stroking swimmer, followed by the middle pair of legs swinging forward to pivot the beetle onward as the front legs reach the backward limit of their arc. The hind legs are largely immobile, acting like sledge runners, gliding the back end of the beetle over the sand.[14] Why? The favourite question of the animal behaviourist. We don't know –a favourite answer. But this offers a chance to find out and thus explain how evolution has created another unique solution to a particular problem. It isn't speed in this case; the galloping beetle covers ground just as fast as when it runs. It could be more efficient in the slipping sand, and it could be a more dependable way of counting steps on a shifting substrate. These beetles, somewhat timid in comparison to their brusque ball-rolling cousins, will eventually share their secret with us.

These secrets, wrested from the descendants of Khepri, have yet to solve our resurrection conundrum. They have made great contributions beyond the thrills of a good wildlife documentary. The way nocturnal insects see in low light has been incorporated into a night vision camera for cars, which allows clear vision in starlight without the need for photomultipliers or infrared light beams. The beetle-brain model might stimulate better design of autonomous drones, or of Global Positioning Systems (GPSs) independent of satellites. By copying the beetles' brain circuitry, low-energy cheap circuits could allow miniaturisation of machines to enter and survey unmapped buildings in search of trapped people. Drones already use insect behavioural algorithms to fly autonomously in uncharted territory. Combining a looming response, where any object rapidly filling the frontal visual field indicates that a collision is imminent, with optic flow – the movement of the visual scene across the periphery of the retina – allows simple robots to fly without colliding with the furniture.

Khepri may not have fulfilled his promises of resurrection, but his worldly namesakes have fertilised not only the earth, but also the minds of scientists eager to understand how life can operate at the base of the food chain, and to learn lessons in both ecological and computational engineering.

CHAPTER SEVEN

Design construction first

WHEN THE EGYPTIANS ELEVATED the humble dung beetle to a symbol of transformation, they were part of a long process of change in the course of human beliefs. We moved from being animists (where each object was imbued with its own spirit) to being theists, where power was held by a limited number of deities and brokered by a priesthood that mediated access to those gods. That journey (from imagining how the earth came into being, to accepting that all life is composed of the same genetic material) has brought us to the point of understanding that we share a common origin with dung beetles and every other living organism on this earth. How that common ancestor has evolved into a myriad different creatures is a fascinating and constantly expanding field of enquiry, and our indomitable little friends are helping us to find the answers to some of those questions.

Dung beetles have become one of the pivotal species' groups in modern evolutionary studies because, small though they are, they are helping us to begin understanding how variation within a species (in horn size, for instance), could promote speciation, which has resulted in so many diverse species in this case. Those 6 000-plus species are found in at least 257 genera (the plural for genus, the next grouping in the classification hierarchy in which species are nested); compare this to humans, where we are only one species (*sapiens*), sitting alone in our solitary genus (*Homo*), with all of our relatives extinct. The transformation dung beetles are now helping us to understand is one of speciation: in other words, how one species can give rise to another.

These recent findings go a long way to resolving Charles Darwin's issues with his own theory of evolution. Darwin had Mendel's discoveries of the mechanisms of inheritance on his own bookshelf, unread; if he had read them, he could have explained the problem of the blending of parental characteristics in their offspring (that over time, this would remove variation from any population, thereby leaving natural selection nothing to work on). In the same way, Darwin came tantalisingly close to an explanation of how new species could arise when he used dung beetle horns as examples of a sexually selected characteristic in *The Descent of Man, and Selection in Relation to Sex* (1871). It is testimony to his talent as a thinker that it was another hundred years before the subject of sexual selection received any serious experimental consideration. Now insects, including dung beetles, provide hard evidence of the validity of Darwin's theory: that new organisms evolve from existing ones.

Possibly the biggest surprise in evolutionary biology (and a blow to our self-importance) which was uncovered in the pursuit of understanding how evolution occurs, was the discovery that basically the same genes regulate the design and function of every organism, including insects and humans. So at first glance, the tool box of our genes is not particularly complicated in its basic form: we share it with some fairly lowly creatures. But the particular combination of those genes, their place in the coding system, and (more elusively) what switches particular genes on and off makes this simple little tool box capable of turning out some very complex creatures.

It was assumed that once the human genome project had been completed in April 2003, we would be able to address the root cause of most diseases, and understand the very essence of life. It has since become apparent, to our frustration (and delight in some quarters, it must be admitted), that genes are not so simple. The tiny and fascinating world of which molecules actually drive all organisms on this planet is more intricate and interconnected than the somewhat narrow belief that reading our genes is all we need to understand life.

Almost a hundred years after Darwin published his ground-breaking study on evolution, James Watson, Francis Crick and Rosalind Franklin unlocked the key to the structure and properties of DNA. Part of their brilliance was their insight that the layout of this long, boring, repetitive molecule could be the gateway to understanding how organisms store their structural information, and how this could be passed from one generation to the next. However, the now familiar notion of the DNA double-helix molecule as a blueprint of our genetic information was in fact not appropriate. A blueprint can easily be recreated by reverse engineering the resulting product. By carefully dismantling a car, blueprints for the component parts can be generated (which can then be used to mass produce Aston Martins in China, for example). Industrial espionage thrives on this phenomenon. But that reversal of information flow is impossible with the DNA of either a human, a chimpanzee or any other organism. The DNA code is more akin to a recipe than a blueprint: a cake cannot be unmade to reveal the recipe, and how the final confection turns out depends as much on how and where it is baked as on its exact ingredients.

The next great hurdle, therefore, after the discovery of the information storage capacity of DNA was to find how it animated, read and distributed that genetic information. In 1961, eight years after Watson, Crick and Franklin's announcement, Marshall Nirenberg and his post-doctoral student Johann Heinrich Matthaei, at the American National Institute of Health (followed by Har Gobind Khorana at the University of Wisconsin, then Robert William Holley at Cornell University) worked out that the code was not read directly from one type of molecule to another. Instead, to make proteins (the stuff of which all biological organisms are built), a process of transcription from DNA into a related molecule (RNA) was followed by a translation into proteins, which are a completely different type of molecule. This sequence: DNA → RNA → Protein was understood to be a strictly one-way flow of information. The environment could select good DNA by dictating which of its protein products (the 'phenotype', which is the living body carrying the DNA) survived or didn't, thereby enabling that DNA (the 'genotype', the package of genes carried by the phenotype) to be passed on to the next generation. At last we had a molecular explanation of how Darwin's natural selection worked: the environment selected the 'fittest' phenotype, which carried the corresponding genotype; these genes were then passed on to the offspring, and used to build the body of the now better adapted phenotype. Most importantly, the environment was considered incapable of speaking back directly to the genotype.

This interpretation of the chain of evolution was so rigidly ensconced in scientific thought that it was called the 'Central Dogma', and briefly hampered the new understanding of the complexity of how life works. We now know (through work on dung beetles and other species) that the phenotype itself can respond to the environment in many subtle ways because the environment can indeed dictate how the genes in the DNA work. The information can and does flow both ways. The Central Dogma got it wrong, and the system is infinitely smarter than we ever imagined.

The field of molecular biology to which Franklin, Watson and Crick belonged was a twentieth century development. It was a world in which matter was broken down into its smallest entities, all of which gave rise to very big questions. Two questions drove this field. One was that of inheritance (which was succumbing to our knowledge of Mendelian genetics), which showed that even if characteristics in the phenotype appeared to have been blended (pink flowers from red and white parents), they could be separated again in the next generation. The second was speciation, which we instinctively knew was true. However, the mechanisms by which a 'hopeful mutant' could spontaneously arise and live to survive and reproduce in all its untested novelty were not clear. How could something truly novel suddenly pop out of a long-standing ancestor, which had earned its spurs by surviving for millennia as a being beautifully adapted to its environment? How could pure chance make that novelty be better than its ancestor?

Darwinian evolution was seen as just tweaking things through the minute gradual changes of micro-evolution, each change perfecting an existing organism, not spitting out a brand-new one from nowhere. This was the other big issue that bedevilled Darwin's theory, and which lay at the core of our understanding of evolution. Darwin knew there had to be something driving the evolution of novelty, from the Last Universal Ancestor (LUA[1]) to the vast array of life on earth. Like his theory of sexual selection, Darwin's theory of natural selection has survived and flourished (through subtle expansion and reinterpretation in Julian Huxley's *Evolution: Modern Synthesis*[2]) to what we have now in the contemporary Extended Evolutionary Synthesis. And here we find that our friends the dung beetles have made their contribution to that development.

Franklin, Crick and Watson's world of molecular biology has a curiously extreme quality worthy of the topsy-turvy world of *Alice in Wonderland*. It is a realm that proves above all that observation and facts are critical to scientific truth, and that truth is frequently the opposite of what was previously thought, or what logic might suggest. For example, the human genome contains far fewer genes than those of mice, or rice; even more amusingly, 75% of our genetic structure is the same as that of a pumpkin.[3] The implications of these odd findings are that neither the complexity of an organism nor its place in the web of life predicts the number of genes it carries, or vice versa. Why this is so is an example of the kinds of questions raised by the apparent disordered nature of genes, which have changed the type of studies in genomics (and the era we call post-genomics) now that the human genome has been published.

Even though the human gene code has been mapped out, it is not possible to understand everything about development of an organism's appearance by looking at genes alone. Because the ways in which those genes are expressed can change, Jean Baptiste Lamarck's notion of epigenetics (where adaptation occurs in an organism through modification of gene expression, rather than alteration of the genetic code itself) has been reintroduced. Notorious for proposing that acquired characteristics could be inherited (passing on to a child the muscles of a body builder or a parent's suntan, for instance), Lamarck's example of the giraffe passing on its stretched neck to its offspring has become a standard schoolroom point of ridicule of the otherwise eminent eighteenth century French scientist. But Lamarck may well have been on the right trail, because there is now increasing evidence that environmentally influenced traits (such as responses to nutritional stress) can be inherited. They remain coded in the DNA, but their expression can change due to a process called methylation, where the addition of a methyl molecule to the DNA can change the activity of a segment without changing the sequence. These traits can therefore be temporarily switched on and off by an environmental experience. However, such changes can only last two generations; so the hardwiring of DNA is not lost; rather, its translation to proteins is temporarily (and reversibly) altered by the environment.

For example, mice carrying the 'Agouti' variant of a gene are genetically identical.[4] However, depending on what their mother ate during pregnancy, the offspring can differ dramatically: they can be brown and skinny with the mutation switched off; they can be fat, yellow and prone to obesity and diabetes when it is on.[5] The switch comes from the mother's environment – which talks back to her genome and influences the fate of her offspring.

So evolution is not a simple one-way street in which an organism's development is driven by its genome, and the phenotype simply reacts to an environment within the limits of its genes. Indeed, the phenotype can extend beyond the body of the organism into the world around it. A common example is the building of dams by beavers – these manifestly alter both the environment and the beavers' success within it. Huge termite mounds are an entomological example of an insect with a phenotype extending well beyond its tiny body. The key question is: where does the variation upon which natural selection needs to act come from?

The extended phenotype theory retains the source of variation within the genome, and argues that these arise as genetic mutations. Meanwhile, proponents of the alternative (but similar-sounding) Extended Evolutionary Synthesis reject the notion that these variations always arise from what are usually random copying errors that happen inside all of us when we replicate our DNA to make new cells. Instead, they suggest that changes in the phenotype can happen before any changes take place in the genotype; this reverses the Central Dogma, allowing the phenotype to talk back to the genes – thereby altering the heritable material. They also interpret most changes in the phenotype as positive rather than neutral or detrimental, so that the hopeful mutant starts off with a better chance in life than as a mash-up of bits of Mum and Dad with an extra toe or eye thrown in somewhere. It seems that no sooner do we unlock one door to the mystery of life than a whole new maze branches off into uncharted territory. It is the very stuff of scientific adventure, and even though contemporary science writing can be difficult for non-scientists to read, it presents a world of never-ending surprise and interest. And all the while, our oblivious dung beetles are playing an important role in exploring many of the questions being raised.

The reason dung beetles are so useful in helping unravel the many puzzles around how life expanded and diversified on our planet is because they are found everywhere except the Antarctic and Greenland. They are relatively abundant, easy to study, and have delightful variation within and between species to act as useful tools for exploring evolution. Many dung beetles are now being studied to answer questions about sexual selection, species formation, hormonal regulation of development, biological control of invasive species and even brain structure, as well as being used in conservation and forensic biology. One species in particular, *Onthophagus taurus*, has been used extensively because many (but critically not all) of the males have a variety of horns. These six-millimetre horns and the shapes and forms that occur within the same species are key to understanding how and why physical changes and novel features can occur almost overnight within a species. Aldrovandi mentioned this bull-headed dung beetle in the sixteenth century, and Fabre dug up its burrows two centuries later. It has remained in the academic spotlight ever since, as a biocontrol agent and now as the lab rat of evolutionary studies. More than 250 scientific papers per year are written on this diminutive creature, which has been introduced most recently to New Zealand, following its successful establishment in Australia and North America in the last century. As a dung beetle 'brand', *Onthophagus taurus* has (almost) the tenacity and reach of Coca Cola (but with rather more beneficial effects).

Darwin suggested that the horns he observed on the beetles he fondly collected were probably an ornamental device to attract females. Alfred Wallace (usually his staunch supporter) disagreed, arguing that they were for survival, for defence against predators. If Darwin was correct, horns would then be a secondary sexual characteristic (not directly involved in survival or the process of sex, but nevertheless influencing a female's choice of mate) and therefore potentially one of the main drivers of natural selection and consequently evolution.

Darwin was partly right, but the world of dung beetles is more aggressive than the Victorian gentleman had imagined. The horns are used for fighting by males who guard tunnels with the intention of controlling sexual access to a female building a nest in that tunnel. However, even when successful in a fight, the horned male might still not be the sole father of her offspring. *Onthophagus taurus* females are free with their favours; more than 80% will mate with at least one other male and five per cent cavort with as many as five sires. But what's in it for the females, other than a good time with the guys? Sex is a risky business for all of us. It can be dangerous – even dung beetles carry sexually transmitted nematodes. Having sex takes time away from other life-sustaining activities like eating and, certainly for the male, requires investment in costly courtship behaviours. If one partner will do, why risk accepting more?

For the male, the answer seems obvious: if more matings translate into more offspring, then he should generate as many copies of his genes as possible in the next generation. But this only works if he is assured of paternity – and we know female dung beetles play the field. To guarantee that his courtship efforts will not be superseded, he must therefore hang around after the deed, doing what is known as mate guarding.[6] Humans will recognise this behaviour in the romantic hand-holding of courting couples, in which a man holds the hand of his partner, thereby physically preventing access to his mate by other prowling males. Dung beetles do something similar: the successful sire sticks around more often if other males are present, and he also indulges in 'insurance copulations'. To be fair, he also helps build brood balls, reducing the cost of parental care for the female and extending her lifespan to boot.

The female dung beetle, on the other hand, can benefit from cheating by setting up competition among her suitors to select the best genes available to her, and therefore her offspring. The only problem with dung beetles is that it is difficult for the females to pick the winners from the losers. Just because her guy won all the fights on the block doesn't necessarily indicate that he comes from a good family. He won because he was big and had horns, because horned males just about always win fights – so horn size is a perfect predictor of fight outcome. So what's the problem? The female just needs to check out the size of a potential mate's horns, and she has good genes in the bag. But not so fast! Horn size in dung beetles has a very low heritability – meaning that a large-horned father has no guarantee of having a large-horned son, and therefore likewise for the female who mates with him. Body size is what dictates horn size, and body size is almost totally driven by the size of the brood ball in which the beetle larva grows. This in turn is open to the vagaries of season, rainfall and dung availability, which can cause good genes to be hidden in a small hornless male whose mother was unable to make a large ball for his larval provision. What is a girl to do? Well, there is one way to sort the wheat from the chaff: set up a lottery.

Sperm competition is rife in dung beetles. Choosing the best male specimen is an impossible task for the females, so instead they set up conditions in which the males' sperm can compete. By mating with multiple males and having a sperm storage organ, female dung beetles ensure that the sexiest (and strongest) sperm will fertilise their eggs. The storage organ, common to most insects, is called the spermatheca. It prevents the successful courter from placing his sperm directly onto the female's eggs, and allows a secret selection process to take place inside her (this is known as 'female cryptic choice'). A female therefore has the last word in who gets to make her babies. From dung beetles to humans, there is no known species among those that have copulatory sex in which the male sex organ can deposit sperm straight onto a female's eggs inside her body. She calls the ultimate shot. *Onthophagus taurus* (who else?) females have a large spermatheca where the most vigorous sperm (sometimes from smaller males) can outrun sluggards from the larger males to fertilise the huge egg as it passes down the female's oviduct.

By accepting several males, including the sneaky copulators who circumvent the hulking brutes at the tunnel entrances by digging side tunnels behind their backs, the female *Onthophagus taurus* increases her chances of getting good genes for her offspring in a secret post-copulatory contest. The craftier males are smaller, and consequently hornless, but have invested more in the size of their testes. This means that having lost the pre-copulatory contest at the front door (in some cases, they do not even bother to compete), they can increase their post-copulatory chances at the back door. Bigger testes means more sperm for the competition that will take place inside the female, and inspection of the offspring of females with multiple mates shows that paternity is indeed skewed towards specific males in her entourage. Female *Onthophagus taurus* lineages that have been kept monogamous for several generations experience an increase in the 'fitness' of their offspring when they are allowed to mate more promiscuously. So however the lottery works, multiple partners clearly benefit the female.

Of course, neither body nor horn size of the offspring is affected by whether the female has had only one or multiple partners.[7] Those characteristics, as we already know, are driven by the amount of the larval provisions. But even where these are ample, a growing body does not have infinite resources to build each component to its maximum size and functional capacity. The developing beetle's body has to decide how the resource investment will be balanced – between eye size, testes size and horn size, for example.

This trade-off between the relative size of horns, testes and other body parts is particularly evident in dung beetles because heredity has so little influence over the size of the horns and other structures. The decision about where to invest food resources takes place during the development of the larva (which, incredibly, somehow knows what its future adult size will be). In other words, the choice to be a big fighter or small sneaker is determined by the larva, a choice that will dictate everything else about the future life and behaviour of that individual.

How does the beetle larva know which developmental route to follow? Again, because dung beetles grow trapped inside a finite food source, scientists have been able to manipulate that store to prove that the trigger is a nutritional one. And because most insects, including dung beetles, have a punctuated development (proceeding through clearly demarcated stages from egg to larva to pupa) we know that the transition from larva to pupa is the critical stage in which insulin (under the influence of food availability) triggers a juvenile hormone to invest resources in a horn rather than testes, for example. Suddenly, a further similarity between ourselves and dung beetles is revealed:

they also use insulin and other hormones as internal messengers for processes involved with digestion and growth. The common ancestor in our genealogies just crept a little closer.

Spurred on by increasing cases of resistance to post-war organic insecticides like DDT, scientists have also discovered that to maintain a healthy insect in its larval stage requires, among other things, a juvenile hormone (JH). When that hormone is withdrawn, the insect is triggered into its next stage of development, when it will pupate and proceed to emerge as a viable, reproductive adult. Good news for both farmers and for environmentalists? To stop a burgeoning population of insect pests in its tracks, just spray on the JH, which (like any hormone) is required only in miniscule quantities. The pest's larvae hang around longer, but get eaten by birds and other predators, while no other animals (other than these insects) are harmed as they would be by the alternative – aerial sprays of insecticide. Critically for the farmer, the insect lifecycle is blocked, and the next generation stalls in its tracks. The pest insect can't multiply into the millions, and another growing season without a plague of pests can be completed.

The immense power of insect juvenile hormones was brought home by an incident in the South African town of Nelspruit where silkworms (traditionally reared each spring by many South African school children) refused to pupate, instead growing into titanic caterpillars consuming ever-increasing quantities of mulberry leaves. The Nelspruit Boy Scouts conducted a survey at local schools that found that the silkworms on the side of town neighbouring the local orange groves were most affected. Citrus farmers in the district had recently switched to JH as a pest control measure in response to increasing levels of insecticide resistance in the red scale insects that were ruining their crops of oranges. The closest orchard involved was at least three kilometres from the town, so spray drift from aerial application of the insecticide could not be blamed. Instead, molecular quantities of the JH (assumed to be adhering to dust particles) were drifting into town and settling onto the mulberry leaves that local children were collecting to feed their silkworms. The hormone had consigned the caterpillars to a Peter Pan limbo until they eventually died of diseases their oversized, youthful bodies were not equipped to deal with.

Given that this potent hormone can mediate a major switch between life stages in an insect, it is not surprising that it is involved in a body shape switch in dung beetles – between a big-bodied horned major male and its alter ego, the small hornless minor male. These males both have the same genes for a horn, and in fact so does the female (that's another story); it just happens to be switched on in the major male and switched off in the minor. The switch point is almost that clear – off below a certain body size (dependent on the species), or on if the brood ball has enough food to generate the bigger body. But above the critical threshold, the horn size correlates with the male's body size in what is known as an allometric relationship. Allometry describes how the size of body structures change in relation to each other. Bigger people tend to have bigger feet so that they don't fall over, whereas human eye size (width rather than length) is relatively consistent across body size, age and race. It doesn't scale according to body size; big and small adults, male or female, have roughly the same sized eyes. Probably the most dramatic example of allometry in humans can be demonstrated like this: grab your left ear with your right hand by stretching your arm across the top of your head. Now get a two-year-old child to do this. They can't – is this because their arms are too short? In fact, their heads are too big. Human children have relatively enormous heads, a fact overlooked by many medieval painters who depicted children as small adults (explaining why Jesus often looks a little spooky snuggled up to his mother). That body allometry changes as we grow, with our heads increasing very little in size while the rest of the body (especially the legs) extends as adulthood is reached. For the dung beetle, the equivalent metamorphosis takes place during the pupal stage, where the fate of the incipient horn is decided.

Hornless female dung beetles actually do have a horn; they just discard it during the pupal stage once they no longer need it. This pupal pronotal 'horn' is present in males and females and lost in both at adulthood.[8] The pupal horn can be the precursor to thoracic horns (on the midsection of the body) in adult males, but not head horns. The function of this structure is to allow the pupa to break open the skull of its former larval stage, permitting expansion of the adult head into its final shape. This skull, more correctly termed the head capsule, is a tough chitinous cranium built to carry the heavy jaws that are needed to chew up the relatively dry dung of the brood ball. Those jaws won't be needed for this task in the adult beetle, as soft mandibles will sort and sieve the tiny, nutritious components in the liquid portion of wet dung instead. The horn in this developmental role is akin to the egg tooth used by bird and reptile hatchlings to break out of their calcified shells. Like the dung beetle horn, it is usually lost before adulthood.

However it is easy to envisage a simple mutation that causes the persistence of the pupal horn in the adult beetle; indeed, a female dung beetle sometimes sports a masculine horn. Mutation in a well-conserved developmental process pops out a novelty, new from old. But for it to be truly re-purposed in the mutant, it must have an immediate selective advantage for that mutant. In this case, Mr Horny wins more fights and gets his gene-set into the next generation using a structure that had been there all the time, though it had been fulfilling another role. Males fighting over tunnels must therefore have preceded horn evolution, with natural selection picking out the winning novelty (and a species evolving in one dramatic step). Looking at the family tree (the phylogeny) of just the genus *Onthophagus* we can see that horns have arisen independently[9] at least nine times, which tells us that the 'mutations' driving adult male horn development are not only common, but have often generated a successful hopeful mutant.

But the female doesn't need the horn as an adult; after all: it has been shown to impede both tunnelling and flight, plus it also trades off against other structures (such as eye size) in her body's contest for resources. How then does she discard this unwanted, costly ornament? She re-absorbs the tissue from the horn using another genetically controlled pathway that evolution occasionally co-opts to make novel phenotypes – programmed cell death.[10] This sinister-sounding process causes cells to commit suicide in a controlled fashion that carves out the final appearance of an organism in the same way that a seamstress snips out and discards the scraps of fabric that will not contribute to the final shape of a garment. For instance, your fingers turning these pages were similarly sculpted while you were an embryo, when cells between your developing fingers deliberately died resulting in you having separated digits.

The alternative is when the now useless characteristic is retained in the adult, but this occurs only if it does not impose some cost on the bearer. Male nipples are a good example here; males have never had any use for nipples (on their own bodies, that is) because they have never in their lives, or their ancestral lives, suckled their young. Instead the nipples are laid down in the early life of the embryo, when it is described as 'indifferent' (to its sex). Once the Y chromosome starts expressing itself, the trajectory towards being a male begins, blocking the development of female organ ducts and co-opting structures such as the clitoris and labia into their male counterparts, the penis and the scrotum respectively. The nipples remain, however, because (even though they have no function in the adult male) they carry no selective cost to their bearer, and it would require another set of mutations and probable metabolic costs to evolve an additional pathway to delete them.

Men have never been considered feminine just because they have nipples, and neither can the minor male *Onthophagus taurus* be considered effeminate because he lacks a horn. Quite the opposite, in fact. Even though the patterns of gene expression in the head region (where the horns will develop) of minor males is most similar to that of females, gene expression in the brains of developing major males matches that of females and is very different to that of the minor males. This probably influences the paternalistic behaviour of major males, who are more inclined to be loyal parents involved in brood care than the rakish minor males who will sneak away from the home fires for sex at every opportunity. Here is another example of a complete switch (this time in behaviour as opposed to morphology), triggered by the environment, that leads to very different lifestyles in the same species carrying the same genes.

While male *Onthophagus taurus* have a set investment pattern, they either do or don't do parental care, depending on their state of horniness; females, in turn, respond to help. When assisted, they do less than they would on their own, and only the largest solitary females can make a brood ball big enough to produce a major male. But working together, a pair of beetles co-ordinate their efforts and, contrary to seemingly cynical scientific predictions, the extra work by one partner leads to increased effort by the other. The pay-off is larger brood balls, which should yield bigger, fitter offspring. This is thought to be the result of an iterated game where one individual initially co-operates and then continues to match its partner's next action in its own subsequent move. Nevertheless, females still do most of the housework, spending 84% of their time on brood care compared to 48% by the males (which is nevertheless impressive by human standards). Not surprisingly, virgin females live longer than their mated sisters. Raising a family is a costly enterprise wherever it is conducted in the biosphere. So our dung beetles once again show how speciation might be prompted by an environmental change, acting on a flexible breeding system to produce a novelty well adapted to its new environment.

This mechanism probably lies behind the change seen in the size and shape of *Onthophagus taurus* beetles shipped to different corners of the planet. True to form, things are bigger in America, including male *Onthophagus taurus* horns. These are bigger for a given body size in North Carolina immigrants compared to Western Australian *Onthophagus taurus* immigrants. More amazingly, the male genitalia have diverged too; not so much in size as in shape.[11] Consistency in size is easy to explain, which despite our human (male?) obsession with the topic, dictates that average is best: an average-sized male organ will fit into more females than the gargantuan phallus of Priapus. The shape story is in fact both racier and more consequential for rapid evolution than a saga of mere dimensions. The dung beetle aedeagus (the equivalent of a penis), which is used to transfer sperm from male to female, is (as a proportion of body size) relatively enormous in many beetles, equal to about one quarter of body length. What is more, it is always rigid. The aedeagus is made of chitin, the same stiff compound that composes the cuticle, with only a small inflatable part, appropriately called the endophallus, which pops out during copulation. The aedeagus only has to be extended out of the male body to be ready for action. But its allometry is flat. As predicted by the 'be average' theory, even as body size and horn size increase there is no corresponding increase in the size of the aedeagus. An over-endowed male would only reduce his chances of getting those (let's face it) rather ridiculous genes into the next generation.

Reluctant as most men are to capitulate to the demolition of another sex myth, what they know in their heart of hearts is that most women like to be wooed. They want to go dancing; they want to be entertained and tickled by witty tales and humour. Dung beetles are no different. Female *Onthophagus taurus* respond to the male's courtship efforts (in this case, his frenetic drumming on her inflexible body surface with his front legs). They also prefer males with high rates of courtship – more drumming. But the endophallus might also play a subtle courtship role. Variation in tiny chitinous plates (called sclerites) on the end of the balloon-like endophallus have been found to influence a male's fertilisation success, suggesting that they play a role in the sexy sperm competition. By helping to persuade the female to use the inseminating male's sperm, these sclerites act as an internal courtship device, which also enhance her chances of getting her genes into the next generation via his son, also likely to be a sexy beetle.

Even though the aedeagus varies little between the American and Australian *Onthophagus taurus*, their specific sclerite structures are rapidly heading off in different directions in each population, which makes it unlikely that they would now be able to recognise each other as potential mates in the unlikely event that they should meet. In other words, a biological control programme may well be responsible for the creation of at least two new species of dung beetles. This divergence has taken less than 50 years[12] which is not even the blink of an eye in evolutionary terms on a planet where life has persisted for more than three billion years and blossomed in the last 500 million. These changes are evolutionarily instantaneous when compared with the ancestral populations of *Onthophagus taurus* in Italy, for example.

It is now becoming clear that genes which have been switched off for one reason (such as those driving horn appearance in hornless males) can sometimes be redirected to create alternative morphological features without a new genetic sequence being necessary. This indicates one way in which novel changes can develop rapidly within species, and explains how some species show sudden adaptation to new environments without any fundamental changes to their DNA. For example, a manipulation of the 'orthodenticle gene' gives our hero *Onthophagus taurus* an eye in the middle of his head. Unfortunately for the rest of us, many of these gene names are unfathomable, or even worse, inside jokes – like 'Groucho', chosen because of the way that mutation modifies a fruit fly's head, giving it fuzzy 'eyebrows'. But scientists, like everyone else, should be allowed some fun. The 'Ken and Barbie' mutation removes the female or male genitalia, just like their doll namesakes, while 'Cheap date' fruit fly mutants are very susceptible to alcohol. However, our increasingly linguistically austere world is in the process expunging humour from the lab, primarily because these mutations are likely also to be discovered in humans – being found to be carrying 'Lunatic fringe' might be disconcerting for the bearer. Nevertheless, there should be a strong case (somewhere) for keeping 'Shaven Baby' and 'Killer of Prune' in place of 'LFNG O-fucosylpeptide 3-beta-N-acetylglucosaminyltransferase'.

The orthodenticle gene is a homeobox gene, possessed by all bilaterally symmetrical creatures (sliced from the head to the groin, we are pretty much a mirror image on each side). Homeobox genes tell the developing embryo where to put things like arms and legs and heads relative to each other, getting the positioning right and maintaining that symmetry. The orthodenticle gene is in charge of the head region in the fruit fly, organising the midline and structures off to either side of it. Shutting that gene down in *Onthophagus taurus* turns off the horn development and substitutes it with a strange Cyclopean third eye, effectively replacing the horn with an equally complex structure. Because this doesn't happen in another beetle species, the flour beetle, it suggests that the orthodenticle gene has acquired a new role in the dung beetle lineages, directing their head and horn formation. Hey presto: new work for old genes.

Dung beetle DNA is clearly helping to answer some of our questions regarding how new species can arise, but it has also posed other questions, the answers to which are not always immediately obvious. Madagascar is one place where research into the range of dung beetles and their colonisation of the island has yielded some intriguing conundrums via analysis of their DNA.[13] Three questions are central to the exploration of dung beetle populations on Madagascar. The first is when and how did they get to the island; the second is when did they start to diverge; and the third is why?

Madagascar broke away from Africa 160 million years ago and from India, Antarctica and Australia approximately 88 million years ago, but their DNA reveals that the dung beetle species endemic to Madagascar are much younger than that, as are most other flora and fauna on the island. Four dung beetle tribes that now occur on Madagascar had ancestors that somehow crossed the Mozambique Channel between 37 and 23 million years ago, with the most recent arrivals turning up between 30 and 15 million years ago. This fits with the timing of the big evolutionary spurt in the main dung donors, the grazing mammals, on the African continent. The oldest Madagascan dung beetle group, the *Canthonini*, apparently spread out across the island about 14 million years ago, while three related generic lineages diverged 5.6, 9.3 and 12.3 million years ago. Two species of the genus

Scarabaeus in Madagascar have been shown by molecular analysis to be sister species to members of the genus from Southern Africa. Genetic analysis suggests that the estimated time of divergence from that shared African/Madagascan origin occurred between 24.15 and 15.8 million years ago[14] – relatively recent events in evolutionary terms.

The flora and fauna of Madagascar have long been recognised as distinct, giving valuable insights into a certain period in the earth's evolution and the unique trajectories that prolonged geographic isolation can lead to. But when what we could call genetic archaeology developed, the curiosity value of the island's creatures increased further. Using a molecular clock (the rate at which certain molecules like proteins or genes mutate over time) to compare the difference in these molecules from two lineages, such as African mainland and Madagascan dung beetles, we can estimate the point in the past at which they diverged from one other.

The chief question concerning Madagascar was this: if the island split from Africa 160 million years ago (which we can date from geological clocks using the rate of decay of radioactive elements), how then did the plants and animals that evolved only 40 million years ago on the island get there? It has been argued that there must have been land bridges between the mainland and Madagascar, which is now 450 km from Africa. Although plausible, there is no physical evidence of such a bridge and, most compelling, none of the larger animals of Africa managed to use such a crossing to establish their home in Madagascar. The current assumption is that the biological colonisation of Madagascar happened via plant rafts; if this is correct, they spawned at least eight colonisation events in the case of the island's dung beetles. The requirement for repeat events somewhat reinforces the idea that they arrived on rafts, as it is more likely that small groups arrived at different times on such rafts. If the idea of plant rafts sounds like evolutionary biologists literally grasping at straws, the experiences of Harvard biologist Jonathan Losos add credibility to the notion.

Losos was surveying anolis lizards on the Caribbean islands to track how the same lineage evolved to cope in different specialised habitats, depending on what island the lizard ended up on, and what competitors were there when it arrived. Related species could evolve, for instance, on one island into a tree crown specialist while on another island its relative became a trunk dweller. Dolefully conducting field work after a hurricane had sucked everything (including all the plants and animals) off one of his tiny island field sites, which were only a few hundred square metres in size, Losos knew that the evolutionary clock had been reset to zero. As he stood, wondering how long it would take to start ticking again, a huge plant raft nudged up onto the shore. Chance already had a suite of species in the starting blocks. Even better, this parcel had its forwarding address in the form of a street sign that had been ripped off the donor island, along with the plants, hundreds of kilometres away. Such chance events run the storyline of evolution differently, depending on what ingredients are brought together, when and where. No wonder rafting has been labelled 'sweepstakes dispersal'.

The dung beetle transfers were possible because Madagascar was not in its current position at that time in geological history. So while the prevailing currents carried animals and plants that had been swept out to sea into oblivion in most cases, a few fortunate passengers were washed onto the island. Dung beetles may have hitched a ride on those plant rafts alongside their food donors, the ancestors of the lemurs, which also only arrived on the island after it had separated from the African continent. When lemurs on the island speciated and adapted to life in different niches, such as eating fruit as opposed to leaves some 30 million years ago, their companion dung beetles did likewise. Madagascar has almost 300 dung beetle species described to date, almost as many as found in Australia (a much larger but relatively younger island) with 96% of the Madagascan species endemic. However only four dung beetle tribes (out of a possible 13) are found in Madagascar, indicating that the window of colonisation opportunity closed as the island shifted to its present position, causing a reverse in the currents that now make the oceanic journey impossible. *Nanos viettei*, a dominant species in the eastern rainforests, is a typical representative of one group of Madagascan dung beetles. Measuring a mere six millimetres in length, it lives for at least two years, but only breeds in its second year – both very unusual traits for small dung beetles. In addition, even though the vigilant males spend hours mate-guarding, they take no part in brood care which yields only one brood ball. These very specific adaptations to a peculiar lifestyle suggest an extreme specialist with limited chances of ultimate survival. Nevertheless, it is the most abundant species in its region and an example of how, given time and isolation, a species can evolve uniquely successful qualities. Unfortunately, with over 80% of its natural habitat – the indigenous forest – gone, these qualities now lead to its vulnerability.

Madagascar is a large island divided by mountains and rivers, and the south-eastern trade winds dominate its climate, delivering monsoon rains to the East coast but tailing off over the central highlands, leaving the South-Western parts of the island arid. This variety of habitats and climatic zones has allowed many dung beetle species to evolve in relative isolation. Nevertheless, various species in the genus *Nanos* will copulate with each other, indicating that they still recognise each other as potential mates. Genetic analysis of the genus supports the possibility of hybrids in their populations, and reinforces the proposal that dung beetles may speciate rapidly under the right circumstances.

One circumstance that has changed in recent history is the Madagascan dung beetle menu. Cattle dung was added to the bill of fare only about 1500 years ago, with the arrival of cattle. Only four species of dung beetles have now expanded their range and food sources to include the now plentiful ungulate dung, which is radically different to the small odiferous lemur pellets that sustained the colonising beetles for so long. Many of those lemur-poo loyalists will also show up at carrion and fish-baited traps, suggesting that they have learned to get by on limited resources over the millennia – but this doesn't explain why they are so fussy about cow dung, after all ... when they're desperate!

Compounding such questions is the knowledge that some species of dung beetles have extended their tastes way beyond dung. *Thorectes lusitanicus*, a Mediterranean dung beetle, eats acorns. It can and does eat dry rabbit pellets, but consuming acorns is its beetle equivalent of protein shakes. Beetles on the acorn diet have greater body mass, bigger ovaries and better resistance to disease and low temperature conditions during their reproductive period from October to December.[15] Given a choice, they will select acorns over dung, with the tree benefitting from the seed dispersal and burial behaviour of the beetles. Dung beetles are well known for inadvertently rolling away and effectively planting seeds that have passed unharmed through the gut of a herbivore. *Thorectes lusitanicus* may have first encountered acorns in this manner, and eventually evolved the ability to use them as a food source without the need for a go-between. However, some plants are more devious in the manner by which they dupe dung beetles into providing courier services. The restios are a rush-like group of plants endemic to South Africa. Typically, their seeds are long and shiny, but one species (*Ceratocaryum argenteum*, in the fynbos Cape floral kingdom) has evolved seeds that look and smell like pellet-shaped antelope droppings. Two Cape species of dung beetles (*Epirinus flagellates* and *Scarabaeus spretus*) have fallen for the trick, dutifully rolling away and burying the seeds – which are too hard for them to even attempt to eat. Unharmed and safe from fire, a fynbos certainty, the buried seeds await the right conditions for germination.

The Cape region in South Africa has another peculiar genus of dung beetles, *Sceliages*, which eat millipedes. Given that as a group, millipede species are poisonous (producing an incredible array of noxious chemicals including alkaloids, quinines, ketones and hydrogen cyanide) and are difficult to roll up and pack into a hole, the novelty is further heightened by knowing that at least one South American species of dung beetle does the same. Except that the South American *Deltochilum valgum* actually kills the chosen millipede, making it (of all things) a predatory dung beetle. Other South American dung beetles prey on leafcutter ants. Two species of ball rollers, *Canthon dives* and *Canthon virens*, gather queens of different leaf-cutting *Atta* species as they innocently emerge from their nest for their nuptial flight, upon which they are unceremoniously rolled up into brood balls for the beetles' larva.[16] These unwise virgins at least get to carry someone else's genes into the next generation.

The apparently aberrant feeding habits of certain dung beetle groups are not so strange if we consider that the subfamily is thought to have evolved from ancestral beetles that fed on humus. Dung beetles are still detritivores when all is said and done, and detritus describes a broad continuum of material from cut grass through to dead (and maybe not so dead) animals and plants, with dung somewhere in the middle. What the beetles actually eat in the largely liquid foods they utilise we still do not know for sure, especially in the few species that eat dry dung (such as the 13 species of *Pachysoma* found on the arid West coast of Southern Africa).

Pachysoma species reveal many other behaviours alien to their closest relatives. Burying and eating dry dung has modified their mouth parts for chewing and not sieving, and their nesting behaviour has meant foregoing the production of a brood ball for their larvae. Instead they make a loose collection of dry detritus, which may include plant material in some species, which is packed into a nesting tunnel. The nest is provisioned by repeated foraging trips in place of the single hit-and-run (or grab-and-roll) strategy typical of their cousins. But, strangest of all, they have given up on flight. That most magical of abilities mastered by so few phyla (birds, bats and insects) has been forsaken in a trade-off we don't quite understand. Dung is usually found in clods and dollops dotted across the landscape, which may be separated by kilometres if the donor is an elephant. Walking between these fragrant oases when one has a stride length measured in millimetres would appear to be evolutionary suicide when one's relatives (and competitors) can fly. The loss of flight therefore must involve a massive gain elsewhere, and a flightless species from the opposite coast of South Africa suggests that saving water is the key.

Every breath of every animal on our dry terrestrial planet is a physiological challenge, balancing oxygen intake and carbon dioxide expulsion along with the regrettable loss of water. We can't help it: our physiology evolved in water, and all our gas exchange trades invoke a water penalty. *Circellium bacchus*, the huge enigmatic beetle that trundles around the elephant reserves of the Eastern Cape of South Africa, is also flightless. It breathes into the space under its wing covers where its wings used to be. This sets up a region of high humidity where the exhaled CO_2 remains for up to one hour, until it is pushed back through the body and out through the breathing holes (spiracles) behind the front legs. Effectively holding its breath for long periods reduces the number of exchanges it has to make with the atmosphere, and presumably lessens water loss.[17] *Circellium bacchus* also apparently eats rat pellets, which are plentiful in the dense bush it shares with the rodents, so the next meal is probably never far away. *Pachysoma* (which dines on dry dung pellets) also don't have to go far for their food, so it might be assumed they would have made the same breathing trade-off with flight, like *Circellium bacchus*. Yet *Pachysoma* breathe just like all their other winged relatives, so another explanation for what at first appeared to be convergent evolution must be found.

Malcolm Coe's Kenyan studies into the dung recycling efforts of dung beetles was the first estimate we had of the enormous job the beetles perform in the African savannah. Subsequent studies have confirmed that they not only remove harmful parasites and pathogens from the soil surface, but they also improve soil tilth, recycle nutrients, and increase water percolation into the earth. They also help to disperse seeds, either willingly or unwittingly. Meanwhile, in Malaysia, two species of dung beetles even pollinate orchids. The beetles are attracted by the rank odour of the orchid flowers, which are attached low on the plant, close to the ground. Once again the poor beetles are deceived as they gain nothing in return for their services, being unable even to eat the pollen they transfer between flowers in their dead-end searches for food. All these jobs now fall under the title of ecosystem services.

Obviously, dung removal is the core business of dung beetles – literally, at the bottom end of the food chain. In its primary form, dung is frequently pathogen-laden and unhealthy, particularly when human dung is found in close proximity to people – this allows parasite lifecycles to be completed as they move from one person to another in easy steps, sneezes or handshakes. Together with earthworms, dung beetles perform Herculean tasks by entombing human dung in the earth. From as early as the 1920s, studies reported that dung beetles in rural India were capable of interring 40 000 – 50 000 tons of human faeces in only two months. Likewise in the rural American South in the 1950s, where there was an absence of proper sanitation, dung beetles performed a vital role in removing human stools from the earth's surface. It was found that, in some instances, it took only 75 minutes for a stool to disappear completely. The dung beetles were not the only creatures involved, but they were by far the largest number and the most active of the consumers and distributors of human faeces. Both rollers and tunnellers got involved in this very public service, with the rollers rolling balls during the day and tunnellers burying dung at night. It was calculated that in an area of about 810 square metres, the dung beetle population was capable of removing the daily excrement of four to six people, or the equivalent of 750 grams of excrement daily.[18]

The majority of such studies are now more often linked to livestock, which says something about improving public health facilities across the globe. Apart from the recycling of valuable elements such as nitrogen, phosphorus and potassium,[19] dung beetles also contribute to the control of greenhouse gases. Methane in particular is produced in abundance in accumulations of animal dung, notoriously generated in feedlots. Pasture-fed animals on the other hand, drop pats that are packed away by beetles with minimal methane release. Dung beetles' role in pathogen control is still under debate; it is not certain whether they prevent the spread of diseases such as hookworm, or transmit bacteria such as *Escherichia coli* (*E. coli*) and *Vibrio cholerae* (*V. cholerae*). Although adult dung beetles are known to destroy many worms' eggs through their peculiar handling of the liquid and fibrous components of dung, not all of the dung (and therefore the parasite eggs) reach the mouthparts in a dung ball – by rolling their dung balls around, the beetles may in fact be spreading rather than containing parasite eggs.

An attempt was made in the US to quantify the value of dung beetles to the American cattle industry. The estimate was based on the figure of the known number of cows in agricultural production – in 2006, this was 100 million cattle. In dung terms, that translated into approximately 900 000 kg of dung per year. Of those 100 million bovines, 56% were treated with avermectin, an anti-parasitical treatment which is toxic to dung beetles. The value estimate was therefore only applied to the 44% of untreated cows. Even so, the financial benefit came in at $0.38 billion per year. [20] Given that this service is performed for free, dung beetles must represent the best value for money to any cattle breeder; a point reinforced by the recent major investment in a new Australian dung beetle programme. Nevertheless, it must be remembered that dung beetles do have friends to help them, and other creatures perform equally valuable tasks in the same department. Ecology cannot be viewed in a single dimension.

Dung beetles belong to what seems to be a trinity of benign creatures whose contribution to the earth and its overall health cannot be overestimated. Termites, earthworms and dung beetles all process waste, transforming it into something life-enhancing. There might well be many other creatures, even the mites that cling to some dung beetles, which perform some sort of ecological task as yet not understood.

Now that we can open up some of the pages of the genetic recipe book for dung beetles, we are at the dawn of an even more exciting period of study. Genetic analysis and comprehension of the workings of evolution appear to be at about the same stage that science was in the eighteenth century, when the global project of naming the world's flora and fauna was just beginning in earnest. We do not know the content of the many unturned pages of the complexity of the natural world, but we do know that it is far more subtle and interconnected than was ever imagined. If one simple creature (whose main activity in life is consuming dung) can be so complex, we can begin to appreciate the distance we still have to travel to unlock millions of years of evolutionary development.

CONCLUSION:

'What a wonderful world'

EVOLUTIONARY THEORY IS MOST commonly described in the catchphrase 'the survival of the fittest', which implies a degree of competition and lack of co-operation. The curious thing about dung beetles is how, in their evolutionary niches, they have both a combative *and* a co-operative existence. On any pile of fresh rhino dung, the array of dung beetles is mind-boggling, from the tiniest ones barely visible to the naked eye, to the large ball rollers. It is the collective activity of the different beetles, all working in their specific niches, which ensures that the large pile of dung is rapidly processed.

Ulisse Aldrovandi thought that insects could be described as atoms, because they were so tiny. Five hundred years later this analogy is most appropriate, as we know from the splitting of the atom quite how powerful very small entities can be. Insects are no different: a plague of locusts can destroy fields of food crops; a colony of termites can consume an entire tree and transform dry earth; a hive of bees can make the difference between a productive and a virtually dead orchard. We also know from the previous chapters what happens when a little settler arrives on a foreign shore and transforms life in the way that the dung beetles did in Australia. As we have previously stated, if there were no dung beetles then there might have been no human race, because the levels of disease and faeces on the planet might have led to our demise before we really got going as a new species.

Together with earthworms and ants, dung beetles represent a trinity of earth transformers. They literally change the earth beneath us, and they do so at absolutely no cost to us. We have only begun to start understanding them, and although less curious minds might think that everything that needs to be said about them has already been covered, there is still so much we do not know. We do not know, for example, if all the species have been accounted for; we do not really know how they came to arrive in places like Madagascar, or indeed in many other countries. We do not know why some species are such generalists while others have very narrow geographic ranges and food choices. We know that one species orientates using the Milky Way, but we do not understand how a brain so small can process or remember such information. We know they are attracted to the smell of dung, but we do not really understand how that works and if that sense switches off when they turn their attention to the visual task of rolling ball.

There are many more questions about dung beetles, and they are all of interest for a variety of reasons. In our history of the development of contemporary science, we have seen it evolve from a belief in magic, to one of stocktaking and empirical observation, to interpretation and deepening levels of sophisticated tunnelling into the smallest known particles. We have gone from myth, symbols, vague observation and interpretation of a world run by the gods, to a world with one God, and then to a world in which the boundaries of religion no longer act as the limit to knowledge.

What drove much exploration was not scientific or natural interest, but the quest for money. Gold (and then trade) became the vehicles for global expansion and settlement; the knowledge we now have of how the world works comes with the recognition that so much of what there is, is threatened by the very pursuit that opened up so much of the world. It is an irony that cannot be lost on us as we look at the growing list of flora and fauna on the brink of destruction and extinction. The relevance of what we still do not know about creatures as small and seemingly insignificant as dung beetles, is that we are beginning to understand what Maria Sibylla Merian showed in her paintings: that the world is deeply and fundamentally interconnected. Evolution is the history of that dynamic process, but evolution has its own timetable, and that is why even though there are creatures that adapt relatively rapidly to change (including our species) there are hundreds of thousands of species that cannot. How do we conserve them and the hidden services they provide for us? Simply by conserving the places where they live. Edward Osborne Wilson's Half-Earth Project is an achievable plan to put aside half the earth, pinpointing the diversity hotspots where the variety of life is highest.[1] Africa and South America, where dung beetle diversity is greatest, still glow brightly on the biodiversity hotspot maps. Devoting a fraction of the money spent refloating the world's economy after the 2008 financial collapse would have bought that land and addressed the challenge; instead only 15% of the target areas are currently protected.

Dung beetles and the knowledge of how they function is a route into understanding some of the processes of evolution: what we know is that it is complex and magnificent, and we are only now at the beginning of some sort of comprehension thereof. The questions of sight and smell might be more fully understood within the new field of quantum biology; we might discover that it is sub-atomic vibrations that lie at the core of smell, rather than interlocking molecules – or it might be a combination of the two. We are at such an early stage in our understanding of the quantum world that without a great deal of experimentation and observation, we cannot begin to understand whether these are in fact the driving forces that explain how creatures with rice-grain sized brains can utilise galaxies for path-finding. The real question is: do we have time? And if we don't, what do we lose?

The short answer is everything. We know that if the foundations of a house are chipped at and destroyed over time, ultimately the building will collapse. The same can be said for the potential loss of insects and other very small creatures. They are literally the foundations of so much of the physical earth, which they redistribute. The apparently destructive locust, for example, is really a redistributive agent of protein. When it dies, it decomposes and disperses all the minerals and food that have gone into its body. It might be of no apparent value to the farmer who has lost his crop, but the earth did not evolve solely for the benefit of humans or human farmers. We have no idea what the effect of increasing light pollution has on animals that have evolved to orientate at night using natural moon and starlight. Drastic differences in light quality and quantity between 'prehistoric' night skies and modern, light polluted skies suggest that the celestial compasses of nocturnal insects should fail to operate in the vicinity of many cities. The obliteration of our heritage in the night sky is thankfully being resisted by Dark Sky Associations[2] and the United Nations.

As we read daily of the hacking to death of another rhino or elephant for horns and tusks, we see one less food source for a host of insect colonies, and also for redistribution networks of everything from seeds to micro-organisms. We understandably focus on the large charismatic animals: the elephants, orangutans and others whose time on this earth seems to be finite. But we don't think much about the little creatures that wriggle and bumble around less visibly, working in tandem with the bigger animals. It is partly a hangover from our hierarchical thinking, partly an increasing disconnect from the natural world. Unfortunately, we are all exhausted by predictions of gloom and the daily avalanche of negative information and news. This might well be why, when the story of dung beetles using the Milky Way to navigate was published; the public was entranced. This showed that we still have the capacity to wonder and be excited by nature, and that alone is a very good reason for continuing to study the natural world.

It is a world of far more complexity and excitement than we suspect, and everything we learn adds to the sense that we are gifted with a world of incredible beauty; a kind of *kunstkammer* beyond our imagining. There is a huge irony in the fact that just as we are starting to have the tools, the people and the framework to begin charting and comprehending the riches of the planet, those riches are being used up like a retiree's financial capital.

When we think of our dung beetles, of their long and fascinating path into our belief systems (and ultimately into our knowledge of evolutionary processes), there is a rich seam of interest one would hardly expect from a creature that spends its life rolling dung around. From the first largely unrecorded studies of those beetles by unknown Egyptians to the many institutions which now study them, we can trace the path of human understanding of our world. As a species, we have reframed our world view many times in the last 7 000 years, and as it has expanded, it has engaged with the tiniest animating forces imaginable. Science is literally on the cusp of a whole new world: one in which complexity theory, chaos theory, and quantum biology (to name but three disciplines) will converge as we try to map the unknown pathways of a world of increasing loss.

Will dung beetles adapt to new territories or food sources if given enough time, and will they have enough time? Research to date suggests that despite their love of dung, they are not creatures that simply migrate and adapt come what may. Countries such as Australia hold some of the keys to understanding this process, but even there the widespread use of veterinary drugs in cattle is affecting the numbers of dung beetles that can thrive alongside their new hosts. Nevertheless, if we want to understand population dynamics and adaptation in foreign environments, we have a natural laboratory in Australia, as well as in Hawai'i. In a different context we might ask whether India, which has the highest rate of open defecation in the world, could adopt a programme to introduce greater numbers of dung beetles to alleviate the problem. But what will be the impact on dung beetles and their global distribution if only certain dung beetles, the successful generalists, are introduced into new territories? Would that have a negative impact on more specialised beetles, and what would the broader ecological implications be? The truth is, we simply do not know.

Playing the evolutionary game – making decisions as to who and what is allowed to multiply and survive, either consciously or unconsciously – is a game that has been spurred on since the age of discovery. We are now custodians of the planet and should take that responsibility seriously. We know that much of what happened as a result of introduced species has been ultimately destructive, so there is little to reassure anyone that further redistribution might not have a similar effect. However, careful and considered introductions (as seen in the biological control of alien weeds, such as the prickly pear cactus) show that we can successfully intervene to manage and conserve whole landscapes impacted by human miscalculations. When we explore the magical world of dung beetle evolution on an isolated island such as Madagascar, we see the full and magnificent complexity of evolution, and the many pathways that creatures both great and small have taken to ensure their survival. This is where we can look to uncover quite how complex and refined the process of millions of years of evolution has been, then use this knowledge to inform management interventions that will be essential elsewhere.

If we need a reminder of how much we do not know, then it is in the study of one little sub-family of beetles, whose seemingly infinite complexity and variety absorb the energies of so many researchers across the globe. Even with that devotion and level of research commitment, we are still missing so many pieces in the story of everything beetle-related; and the further we study, the more varied and remarkable they become. Amazingly, we don't actually yet know what dung beetles eat; which portion of the dung is consumed and how they do it. The role of their gut microbiome in this process is largely a mystery,[3] even as we explore the effect of our own on our health and well-being. We have the places, the tools, the people, and the passion, but we don't have the time or the resources. This is only in the world of dung beetles: it is the same for countless other species.

We hope, however, that the glimpse we have provided into the world of dung beetles and their place in our own history demonstrates that the rich tapestry of life (which we want to understand and nurture) is worth preserving so that our collective survival will be a matter for celebration and further wonder rather than despair.

Appendix A

Footnote 4 to Chapter 21 in Darwin's *Journal of researches into the natural history and geology of the countries visited during the voyage of H.M.S. Beagle round the world in which he makes reference to several species of dung beetles, along with his thoughts on how some species had arrived in particular places, while they were notably absent from others.*

Among these few insects, I was surprised to find a small Aphodius (nov. spec.) and an Oryctes, both extremely numerous under dung. When the island was discovered it certainly possessed no quadruped, excepting perhaps a mouse: it becomes, therefore, a difficult point to ascertain, whether these stercovorous insects have since been imported by accident, or if aborigines, on what food they formerly subsisted. On the banks of the Plata, where, from the vast number of cattle and horses, the fine plains of turf are richly manured, it is vain to seek the many kinds of dung-feeding beetles, which occur so abundantly in Europe. I observed only an Oryctes (the insects of this genus in Europe generally feed on decayed vegetable matter) and two species of Phanæus, common in such situations. On the opposite side of the Cordillera in Chiloe, another species of Phanæus is exceedingly abundant, and it buries the dung of the cattle in large earthen balls beneath the ground. There is reason to believe that the genus Phanæus, before the introduction of cattle, acted as scavengers to man. In Europe, beetles, which find support in the matter which has already contributed towards the life of other and larger animals, are so numerous, that there must be considerably more than one hundred different species. Considering this, and observing what a quantity of food of this kind is lost on the plains of La Plata, I imagined I saw an instance where man had disturbed that chain, by which so many animals are linked together in their native country. In Van Diemen's Land, however, I found four species of Onthophagus, two of Aphodius, and one of a third genus, very abundant under the dung of cows; yet these latter animals had been then introduced only thirty-three years. Previously to that time, the Kangaroo and some other small animals were the only quadrupeds; and their dung is of a very different quality from that of their successors introduced by man. In England the greater number of stercovorous beetles are confined in their appetites; that is, they do not depend indifferently on any quadruped for the means of subsistence. The change, therefore, in habits, which must have taken place in Van Diemen's Land, is highly remarkable. I am indebted to the Rev. F. W. Hope, who, I hope, will permit me to call him my master in Entomology, for giving me the names of the foregoing insects.

Appendix B

Dung beetle species established in Hawai'i

Species	Dung handling†	Origin	Year First Released
Ateuchus lecontei (Harold)*	Tunneller	USA	1963
Canthon humectis (Say)	Roller	Americas	1923
Canthon indigaceus Leconte	Roller	Americas	1954
Canthon pilularius (Linnaeus)	Roller	USA	1963
Copris incertus Say	Tunneller	Americas	1922
Euoniticellus africanus (Harold)	Tunneller	Africa	1974
Euoniticellus intermedius (Reiche)	Tunneller	Africa	1974
Oniticellus cinctus (Fabricius)	Dweller	Asia	1957
Oniticellus (=Liatongus) militaris (Laporte)	Dweller	Africa	1957
Onitis alexis Klug	Tunneller	Africa	1973

Onitis vanderkelleni Lansberge	Tunneller	Africa	1976
Onthophagus binodis Thunberg	Tunneller	Africa	1973
Onthophagus gazella (Fabricius)	Tunneller	Africa	1957
Onthophagus incensus (Say)	Tunneller	Americas	1923
Onthophagus nigriventris D'Orbigny	Tunneller	Africa	1975
Onthophagus oklahomensis Brown	Tunneller	USA	1963
Onthophagus sagittarius (Fabricius)	Tunneller	Asia	1957

* The authority who described the species is given as a surname after the species name. If a species is transferred to a genus other than the one in which it was originally described, the author's name is placed in brackets. Clearly dung beetle names keep changing.

† Depending on the species, dung beetles either roll dung balls, pack dung in tunnels under the pat, or conduct their life within the pat.

Notes

Chapter One: When the dung beetle wore golden shoes

1 Brett C. Ratcliffe, 'Scarab Beetles in Human Culture,' *The Coleopterists' Society 60*, sp5 (2006), 99.

2 Lü Dongbin, *The Secret of the Golden Flower: A Chinese Book of Life*, trans. Cleary Thomas (New York: Harper One, 1991), 16.

3 In modern Mexico, the word 'mayate' (which means dung beetle in Nahuatl, an Uto-Aztecan language) is used in city patois to denote homosexuals and people of African descent. It is used in a perjorative sense to refer to groups seen as the underdogs/underbeetles.

4 Yuval N. Harari, *Homo Deus: A Brief History of Tomorrow* (London: Vintage, 2016), 105–112.

5 One of the first written records we have of Khepri is in the Pyramid Texts, which date back as far as 2300 BCE. The inscriptions call for Khepri to come into being as a manifestation of Re (or Ra), the sun god.

6 Isaac Meyer, *Scarabs: The History, Manufacture and Religious Symbolism of the Scarabaeus in Ancient Egypt, Phoenicia, Sardinia, Etruria. Also Remarks on the Learning, Philosophy, Arts, Ethics of the Ancient Egyptians, Phoenicians, Etc.* (New York: Edwin W. Dayton, 1894), 79–80.

7 W. M. Flinders Petrie, *Scarabs and Cylinders with Names* (London: Constable & Co Ltd., 1917), 2–3.

8 D. C. Watts, *Elseviers Dictionary of Plant Lore* (Amsterdam: Elsevier, 2007), 264.

9 The impressive statue at Karnak in Egypt is still reputed to be a place where one goes to seek good fortune, particularly for love and for conception; as recently as the nineteenth century, Egyptian women would eat one of the scarab family, *Ateuchus aegyptiorum*, in order to enhance their fertility.

Ateuchus aegyptiorum has also been called *Scarabaeus aegyptiorum* and is currently known as *Kheper aegyptiorum* (Latreille, 1823).

10 The taxon *Kheper* has had a chequered history in dung beetle taxonomy (the science of ascribing names to species). Noble origins notwithstanding, *Kheper* has been erected and subsumed as a viable genus several times in the last 150 years. This highlights the unnatural nature of the artificial groupings created in taxonomy. Only the living species themselves recognise other members of the same species as viable mates in nature, and even then, mistakes are occasionally made.

11 The full text of this stela can be found on the British Museum's online collection at www.britishmuseum.org/research/collection-online/collection -object-details.aspx?ld=119746&partld=1.

12 Yves Cambefort, 'Beetles as Religious Symbols,' *Cultural Entomology Digest* 2 (1994), 4. https://www.insects.orkin.com/ced/issue-2/beetles-as-religious -symbols/.

13 The Egyptians believed that *Scarabaeus sacer* had thirty toes, which they saw as a further correspondence of his significance to the cycle of life.

14 Alexander Turner Cory, *The Hieroglyphics of Horapollo Nilous* (London: William Pickering, 1840), 20.

15 Jack M. Sasson, *Civilizations of the Ancient Near East*, Vol 3 (New York: Charles Scribner and Sons, 1995), 161.

16 Alexander Ahrens, 'The Scarabs from the Ninkarrak Temple Cache at Tell A Saka/ Terqa (Syria): History, Archaeological Context and Chronology,' *International Journal for Egyptian Archaeology and Related Disciplines* 20 (2010), 431–444.

17 Aristophanes, *The Complete Greek Drama*, Vol 2, trans. Eugene O'Neill, Jr. (New York: Random House, 1938), 118.

18 John Sutherland and Stephen Fender, *Love, Sex, Death and Words: Surprising Tales from a Year in Literature* (London: Icon Books, 2010), 376–378.

Chapter Two: Crawling out of the darkness

1 Many magical practices were banned by the authorities in both the Greek and Roman Empires. This in fact enhanced the idea that there was particular potency and hidden truth in magic.

2   Aristotle, *Generation of Animals*, trans. A. L. Peck (London: William Heinemann Ltd, 1902), 181.

3   Isaac Meyer, *Scarabs: The History, Manufacture and Religious Symbolism of the Scarabaeus in Ancient Egypt, Phoenicia, Sardinia, Etruria. Also Remarks on the Learning, Philosophy, Arts, Ethics of the Ancient Egyptians, Phoenicians, Etc* (New York: Edwin W. Dayton, 1894), 63.

4   James Fujitani, *Simple Hearts: Animals and the Religious Crisis of the Sixteenth Century* (PhD diss., University of California, 2007), 27.

5   Ulisse Aldrovandi, *De animalibus insectis libri septem, cum singulorum iconibus ad viuum expressis* (Bologna, 1602).

6   George Ripley, *Compound of Alchemy. Liber Secretissmus*. (California: English Grand Lodge, Rosicrucian Order, AMORC, 1993 and 2016), 41.

7   Anthony C. Grayling, *The Age of Genius: The Seventeenth Century and the Birth of the Modern Mind* (London: Bloomsbury, 2016), 143.

8   Erik Hornung, *The Secret Lore of Egypt: Its Impact on the West* (New York: Cornell University Press, 2001).

9   The Hermetic texts were eventually shown by Isaac Casaubon (1550–1614) to be the work of a single author, using second-century Greek, but this has not discouraged the determined Occultist or Cabalist over the last five hundred years.

10 Also known as the Bembine Tablet; it is preserved in the Turin Museum.

11 Kircher believed that 'panspermia' was a vital force of nature used by God to transform primordial chaos into the ordered universe.

12 Paula Findlen, *Athanasius Kircher: The Last Man Who Knew Everything* (New York: Taylor and Francis Books, 2004), 200.

13 The Polyandria were a botanical class of plants whose flowers had twenty or more stamens; now an obsolete description.

14 Joan Ward-Harris, *More Than Meets The Eye: The Life and Lore of Western Wildflowers*. (Oxford: Oxford University Press, 1983), xviii.

15 Tony Griffiths. *Stockholm: A Cultural and Literary history*. (Oxford: Oxford University Press, 2009), 129.

16 The complicated transposing of the stag beetle into a symbol of Christ derived from its mandibles being likened to the antlers of deer that reputedly could

fight off snakes (the devil). The deer was also a symbol of Christ, and by association, the stag beetle became a symbol of Christ.

17 Eric Jorink, "'Outside God, There is Nothing": Swammerdam, Spinoza and the Janus-Face of the Early Dutch Enlightenment,' in *The Early Enlightenment in the Dutch Republic, 1650-1750: Selected Papers of a Conference held at the Herzog August Bibliothek,* Wolfenbüttel 22–23 March 2001, ed. Wiep van Bunge (Leiden: Brill, 2003), 82.

18 Eric Jorink, 'Between Emblematics and the "Argument from Design": The Representation of Insects in the Dutch Republic,' *Early Modern Zoology: The Construction of Animals in Science, Literature and the Visual Arts* (2 Vols), ed. Karl A. E. Enenkel and Mark S. Smith (Leiden: Brill, 2007), 147.

19 In 1758 Rösel published a book on frogs: *Historia naturalis Ranarum nostratum.*

20 John Brickell, John Lawson and John Bryan Grimes, *The Natural History of North Carolina: With an Account of the Trade, Manners* (Dublin: James Carson, 1737), 11.

21 Brickell, Lawson and Grimes, *The Natural History of North Carolina,* 162.

22 Pehr Kalm, *Travels into North America: Containing its natural history, and a circumstantial account of its plantations and agriculture in general, with the civil, ecclesiastical and commercial state of the country, the manners of the inhabitants, and several curious and important remarks on various subjects,* Vol 1 (London: T. Lowndes, 1770), 4–5.

Chapter Three: Joining the dots

1 Paul Lawrence Farber, *Finding Order in Nature: The Naturalist Tradition from Linnaeus to E. O. Wilson* (Baltimore: John Hopkins University Press, 2000), 6.

2 Theodore D. A. Cockerell, 'Dru Drury, an Eighteenth Century Entomologist,' *The Scientific Monthly* 14, 1 (1922), 76.

3 Cockerell, 'Dru Drury, an Eighteenth Century Entomologist,' 78.

4 Cockerell, 'Dru Drury, an Eighteenth Century Entomologist,' 78.

5 Cockerell, 'Dru Drury, an Eighteenth Century Entomologist,' 72.

6 Sören Ludvig Tuxen, 'The Entomologist, J. C. Fabricius,' *Annual Review of Entomology* 12, 1 (1967), 1–15. https://www.annualreviews.org/doi/pdf/10.1146/annurev.en.12.010167.000245.

7   Warren R. Dawson, *Catalogue of the Manuscripts in the Library of the Linnean Society – Part I. The Smith Papers: The Correspondence and Miscellaneous Papers of Sir James Edward Smith M.D., F.R.S., First President of the Society* (London: Linnean Society, 1934), 59.

8   Paul White, 'The Purchase of Knowledge: James Edward Smith and the Linnaean Collections,' *Endeavour* 23, 3 (1999), 126–129.

9   S. L. Tuxen, 'The Entomologist, J. C. Fabricius,' 6–7.

10 S. L. Tuxen, 'The Entomologist, J. C. Fabricius,' 12.

11 Kenneth G. V. Smith, 'Darwin's Insects: Charles Darwin's Entomological Notes,' *Bulletin of the British Museum (Natural History)* 14, 1 (1987), 48.

12 Smith, 'Darwin's Insects: Charles Darwin's Entomological Notes,' 62.

13 Richard Keynes, *Charles Darwin's Zoology Notes & Specimen Lists from H.M.S. Beagle* (Cambridge: Cambridge University Press, 2000), 175.

14 Keynes, *Charles Darwin's Zoology Notes & Specimen Lists from H.M.S. Beagle*, 234.

15 Smith, 'Darwin's Insects: Charles Darwin's Entomological Notes,' 81.

16 Charles R. Darwin, *The Descent of Man, and Selection in Relation to Sex* (London: Murray, 1890), 297.

17 Douglas J. Emlen, 'The Evolution of Animal Weapons,' *Annual Review of Ecology, Evolution, and Systematics* 39 (2008), 387–413.

18 John George Wood, *Insects Abroad: Being a Popular Account of Foreign Insects Structure, Habits and Transformation* (London: Longman, Green and Co., 1892), Preface. https://www.biodiversitylibrary.org/item/62694#page/9/mode/1up.

19 'Science: A Four Thousand Year History,' *Soapbox Science Editor*, last modified March 30, 2011, http://blogs.nature.com/soapboxscience/2011/03/30/science-a-four-thousand-year-history-1.

20 Augustin Fabre, *The Life of Jean Henri Fabre*, trans. Bernard Miall (New York: Dodd, Mead and Company, 1921), 392.

21 Fabre, *The Life of Jean Henri Fabre*, 392.

22 Fabre, *The Life of Jean Henri Fabre*, 107.

23 Fabre, *The Life of Jean Henri Fabre*, 108.

24 Fabre, *The Life of Jean Henri Fabre*, 25–26.

25 Jean-Henri Fabre, *The Sacred Beetle and Others* (London: Hodder & Stoughton, 1981), 186.

26 Mario Livio, *Brilliant Blunders: From Darwin to Einstein – Colossal Mistakes by Great Scientists that Changed our Understanding of Life and the Universe* (New York: Simon and Shuster, 2013).

Chapter Four: Colonising insects

1   Sydney Ross, 'Scientist: The Story of a Word,' *Annals of Science* 18, 2 (1962): 65-85. https:doi.org/10.1080/00033796200202722.

2   B. J. Walsh, 'The Texas Cabbage Bug,' *Practical Entomologist* (1866), 1–10.

3   Even though the mynah had been a disaster in Hawai'i, it did not prevent the ill-advised introduction of the mynah to the island of Mauritius to control insects in sugarcane. Repeated introductions of the Indian mongoose to fragile island ecosystems around the world continued through to the 1950s (when it was introduced to the Comoros islands) and then it found its way to Hong Kong by 1989.

4   Also known as the cottony cushion scale, *Icerya purchasi* is a dramatic beast – for a scale insect. Like many other scale insects, the parthenogenic females protect themselves by living beneath a wax coating, which gums up the mouthparts of their predators. The wax of this scale is, however, neatly extruded in pleated folds, giving it the appearance of a grooved button. True to form as a Victorian hitchhiker, it is now a cosmopolitan pest of more than 50 tree species worldwide.

5   The Vedalia beetle, *Rodolia cardinalis*, along with the parasitic fly, *Cryptochaetum iceryae*, now keep the fluted scale under complete control. Koebele's 514 Vedalia beetles, which he imported over the winter of 1888 CE–1889 CE, proliferated so rapidly that by 12 June, 10 555 were redistributed to 208 growers around California. The beetle is used around the world to control this scale insect.

6   Rolla P. Currie and Andrew N. Caudell, 'An Index to Circulars 1–100 of the Bureau of Entomology,' *U.S. Department of Agriculture, Bureau of Entomology, Circular 100.* (Washington: The Bureau 1911), 235–6.

7   A parasitoid is a parasite that completes its development in or on one host and is usually responsible for the death of that host, as opposed to parasites which do not usually kill their hosts. Many fly and wasp species are parasitoids and therefore important for pest control.

8 Currie and Caudell, 'An Index to Circulars 1–100 of the Bureau of Entomology, 235–6.

9 The Allee Effect describes how an individual's chance of mating declines to such levels at a low population size that it can result in a population crash to extinction. Many endangered species face this threat. The highly gregarious passenger pigeon may well have succumbed via this mechanism.

10 Jennifer Jacobs et al., 'First come, first serve: "Sit and Wait" behaviour in dung beetles at the source of primate dung,' *Neotropical Entomology* 37, 6 (2008), 641-645.

11 In contrast to Darwin's passion for scarabs, it is remarkable that in the 488 pages of *The Malay Archipelago*, Wallace never once specifically mentions any dung beetles, which he must have encountered during his eight years of travelling, or found among the 83 200 specimens of beetles he brought back to Britain.

12 Frank W. Nicholas and Jan Nicholas, *Charles Darwin in Australia* (Cambridge: Cambridge University Press, 2008), 155.

13 Subsequent experiments on the gut microbiota of adult and larval dung beetles copied the formalin washing method used by the CSIRO, but found that bacteria could be cultured from the washed eggs. A biofilm on the egg surface first has to be stripped off with alcohol to expose the micro-organisms to a bleach wash in order to render the egg surface truly sterile.

Chapter Five: Of elephants and dung beetles

1 Paul Hermann Muller, a Swiss chemist, was awarded the Nobel Prize for its discovery.

2 John Frederick Walker, *Ivory's Ghosts: The White Gold of History and the Fate of Elephants* (New York: Grove Press, 2010), 106.

3 The Columbian exchange, named after Christopher Columbus, refers to the intentional and unintentional consequences of the movement of people, flora, fauna, diseases and commodities between continents during the era of colonisation. Smallpox, for example, was intentionally introduced by the Spanish to the Americas on infected blankets. It had a devastating effect on indigenous populations had no immunity against it.

4 Malcolm Coe, 'Dung: The Natural History of the Unmentionable,' *SWARA*, East Africa Wild Life Society Vol 2, No 3 (1979), 8.

5 Olive Schreiner, *The Story of an African Farm*, ed. Patricia O'Neill (Ontario: Broadview Press, 2003), 112.

6 The hister beetles are shiny black beetles often associated with dung. They usually hunt fly larvae in dung and other rotting detritus.

7 Now renamed *Digitonthophagus gazella*. Even entomologists get fed up with having to learn new names for old friends. However, the process is necessary to keep order as the world's (dwindling) catalogue of species is continually reshuffled and we unravel new and often more accurate relationships between those species.

8 Cataloguing collections is a demanding task, as the expert involved has to compare the species being named with all the possible candidates for that name, which may have been described elsewhere in the world. The 'type' specimens associated with that species' name often lie in massive collections in Paris and London, or more difficult to access museum collections. Before the advent of the internet, the type specimens, or paratypes, would have to be requested by post from the holding collection, with the result that entomological egos could hold specimens and consequently careers to ransom. Nowadays, a series of high-definition photographs (coupled with internet-accessed descriptions) make the taxonomist's task faster, less stressful and more egalitarian.

9 This is according to Scopus (a research database) figures for numbers of research papers published in a given period. Papers on dung beetles from 1960 to 1986 = 125 papers; 1987 to 2017 = 1356 papers; papers on *Drosophila* from 1960 to 1986 = 14 492 papers; 1987 to 2017 = 80 561 papers.

Chapter Six: Tribes with human attributes

1 Citation for the Badzon Prize for Biology, which von Frisch won in 1962.

2 Bernd Heinrich and George A. Bartholomew, 'Roles of Endothermy and Size in Inter- and Intraspecific Competition for Elephant Dung in an African Dung Beetle, *Scarabaeus laevistriatus*,' *Physiological Zoology* 52, 4 (1979), 484–496.

3 Jochen Smolka et al., 'Dung Beetles Use their Dung Ball as a Mobile Thermal Refuge,' *Current Biology* 22 (2012), R863–R864.

4   Geoffrey. D. Tribe, 'Pheromone Release by Dung Beetles (Coleoptera: Scarabaeidae),' *South African Journal of Science* 71 (1975), 277–278.

5   Barend. V. Burger et al., 'Composition of the Heterogeneous Sex Attracting Secretion of the Dung Beetle, *Kheper lamarcki*,' *Z. Naturforsch* 38 (1983), 848–855.

6   Chevonne Reynolds and Marcus J. Byrne, 'Alternate Reproductive Tactics in an African Dung Beetle, *Circellium bacchus* (Scarabeidae),' *Journal of Insect Behaviour* 26 (2013), 440–452. 10.1007/s10905-012-9365-1.

7   Douglas J. Emlen, 'Costs and the Diversification of Exaggerated Animal Structures,' *Science* 291 (2001), 1534–1536.

8   Marcus J. Byrne, Bronwyn Watkins and Gustav Bouwer, 'Do Dung Beetle Larvae Need Microbial Symbionts From Their Parents to Feed on Dung,?' *Ecological Entomology* 38 (2013), 250–257. 10.1111/een.12011.

9   Marcus Byrne et al., 'Visual Cues Used by Ball-Rolling Dung Beetles for Orientation,' *Journal of Comparative Physiology* A 189, 4 (2003), 411–418.

10   Marie Dacke et al., 'Animal Behaviour: Insect Orientation to Polarized Moonlight,' *Nature* 424, 1 (2003), 33.

11   Marie Dacke et al., 'Dung Beetles Use the Milky Way for Orientation,' *Current Biology* 23 (2013), 1–3. http://dx.doi.org/10.1016/j.cub.2012. 12.034.

12   Nathan P. Lord et al., 'A cure for the blues: opsin dedication and subfunctionalization for short - wavelength sensitivity in jewel beetles (Coleoptera: Buprestidae),' *BMC Evolutionary Biology* 16(1), 2016

13   Esa-Ville Immonen et al., 'Anatomical organization of the brain of a diurnal and a nocturnal dung beetle,' *Journal of Comparative Neurology* 525, 8 (2017), 1879–1908.

14   Jochen Smolka et al., 'The Galloping Dung Beetle: A New Gait in Insects,' *Current Biology* 23, 20 (2013), R913–915.

Chapter Seven: Design construction first

1   The last universal ancestor (LUA) (also called the last universal common ancestor, LUCA) – that is, the most recent common ancestor of all currently living organisms – is believed to have appeared about 3,9 billion years ago.

2   Julian Huxley published his book *Evolution: Modern Synthesis* in 1942 summing up all that was known up to that date in evolutionary

biology. The book's title became the name that described the field for the rest of the twentieth century. Essentially, the synthesis blended the work of Darwin and Mendel to explain the evolution of species, but it was flawed for a number of reasons. It worked on the assumption that evolutionary changes happened gradually (they don't), and it extrapolated from micro to macro, while subsequent research has revealed that life is more complex than a simple relationship between genotype and phenotype.

3   The human genome contains 20 000 to 25 000 genes, the domestic cat has 20 285 genes, the mouse 24 174, and rice 32 000 to 50 000.

4   'Agouti' is an unfortunate biological insider term. Agoutis are rodents, like mice, but more closely related to guinea pigs. Their hairs have alternate bands of colour, giving them a salt-and-pepper appearance that agouti mice share.

5   The yellow agouti mice reveal remarkable differences in hair colour and weight, influenced by epigenetic modifications. Mice whose agouti gene is 'on' are more likely to suffer from diabetes and cancer as adults. Furthermore, these epigenetic modifications can be inherited by their offspring.

6   Damselfly males take mate guarding to extremes. They can be seen flying over ponds in tandem, the male unceremoniously gripping the female behind her head. His behaviour gets still more chauvinist: he is able to use his genitalia to scoop sperm out of the female, removing the efforts of any preceding males.

7   Leigh W. Simmons and Francisco García-González, 'Evolutionary Reduction in Testes Size and Competitive Fertilization Success in Response to the Experimental Removal of Sexual Selection in Dung Beetles,' Evolution 62, 10 (2008), 2580–91. https://kopernio.com/viewer?doi=10.1111/j.1558-5646.2008.00479.x&route=6Smith.

8   Insects have three body sections: head, thorax and abdomen. The thorax is composed of three segments, the pro-meso-and metathorax. In beetles, the dorsal surface of the prothorax covers the two other segments to form a curved plate called the pronotum. Many beetles develop horns on that structure.

9   The phylogeny of any organism is like a family tree that stretches back over thousands or even millions of years. Moving forward along the tree to where it branches into two lineages, these new lines usually represent a speciation event, in which one species splits into two. These related species would be

expected to have shared characteristics (like a horn), if their common ancestor had one. However, if only one new lineage carries the horn and the ancestor does not, then we can be certain that the new characteristic has arisen independently in that new lineage. This has happened at least nine times in the phylogeny of *Onthophagus*.

10 Armin P. Moczek, 'Evolution and development: *Onthophagus* beetles and the evolutionary development genetics of innovation, allometry, and plasticity,' *Ecology and Evolution of Dung Beetles*, Leigh Simmons and James Ridsdill-Smith, (Oxford: Blackwell Publishing, 2011), 135.

11 Astrid Pizzo et al., 'Rapid Shape Divergences Between Natural and Introduced Populations of a Horned Beetle Partly Mirror Divergences Between Species,' *Evolution and Development* 10, 2 (2008), 166–175.

12 Armin P. Moczek, 'Phenotypic Plasticity and Diversity in Insects,' *Philosophical Transactions of the Royal Society B: Biological Sciences* 365, 1540 (2010), 593–603.

13 Much of this work was initiated by Ilkka Hanski, who was Malcolm Coe's doctoral student.

14 Heidi Viljanen et al., 'Structure of local communities of endemic dung beetles in Madagascar,' *The Journal of Tropical Ecology* 26 (2010): 481–496.

15 José R. Verdú, et al., 'Dung Beetles Eat Acorns to Increase their Ovarian Development and Thermal Tolerance,' *PLoS ONE* 5, 4 (2010): 1–8. https://doi.org/10.1371/journal.pone.0010114.

16 Luiz Carlos Forti et al., 'Predatory Behavior of *Canthon Virens* (Coleoptera: Scarabaeidae): A Predator of Leafcutter Ants,' *Psyche: A Journal of Entomology* Vol 2012 (2012). https://kopernio.com/viewer?doi=10.1155/2012/921465&route=7.

17 Marcus J. Byrne and Frances D. Duncan, 'The Role of the Subelytral Spiracles in Respiration in the Flightless Dung Beetle, *Circellium bacchus*,' *Journal of Experimental Biology* 206 (2003), 1309–1318.

18 Albert Miller, 'Dung Beetles (Coleoptera, Scarabaedae) and Other Insects in Relation to Human Faeces in a Hookworm Area of Southern Georgia,' *The American Society of Tropical Medicine and Hygiene* 3, 2 (1954), 372–389.

19 Daigo Yamada et al., 'Effect of Tunneler Dung Beetles on Cattle Dung Decomposition, Soil Nutrients and Herbage Growth,' *Japanese Society of*

Grassland Science 53, 2 (2007), 121–129. https://doi.org/10.1111/j.1744-697X.2007.00082.x.

20 John. E Losey and Mace Vaughan, 'The Economic Value of Ecological Services Provided by Insects', *American Institute of Biological Sciences* 56, 4 (2006), 311–323.

Conclusion: 'What a wonderful world'

1 Half-Earth Project. http://www.half-earthproject.org/.

2 http://darksky.org/light-pollution/night-sky-heritage/.

3 Shantanu P. Shukla et al., 'Gut microbiota of dung beetles correspond to dietary specializations of adults and larvae', *Molecular Ecology* 25, 24 (2016), 6092–6106. http://dx.doi.org/10.1111/mec.13901.

Select Bibliography

Ahrens, Alexander. 'The Scarabs from the Ninkarrak Temple Cache
at Tell A Saka/Terqa (Syria): History, Archaeological Context and
Chronology.' *International Journal for Egyptian Archaeology and Related
Disciplines* 20 (2010): 431–444.

Aldrovandi, Ulisse. *De animalibus insectis libri septem, cum singulorum
iconibus ad viuum expressis.* Bologna, 1602.

Aristophanes. *The Complete Greek Drama*, Vol 2, trans. Eugene O'Neill, Jr.
New York: Random House, 1938.

Aristotle. *Generation of Animals*, trans. A. L Peck. Cambridge: Harvard
University Press, 1943.

Brickell, John, John Lawson and John Bryan Grimes. *The Natural History
of North Carolina: With an Account of the Trade, Manners.* Dublin:
James Carson, 1737.

Budge, E.A. Wallis (ed) *Syrian Anatomy, Pathology and Therapeutics or
"The Book of Medicines."* London: Oxford University Press, 1913.

Burger, B. V., Zenda Munro, Marina Röth, H. S. C. Spies, Verona Truter,
G. D. Tribe and R. M. Crewe. 'Composition of the Heterogeneous
Sex Attracting Secretion of the Dung Beetle, *Kheper lamarcki*'.
Z. Naturforsch 38 (1983): 848–855.

Byrne, Marcus J. Bronwyn Watkins and Gustav Bouwer. 'Do Dung
Beetle Larvae Need Microbial Symbionts from Their Parents to
Feed on Dung?' *Ecological Entomology* 38 (2013): 250–257: 10.1111/
een.12011.

Byrne, Marcus J. and Frances D. Duncan. 'The Role of the Subelytral
Spiracles in Respiration in the Flightless Dung Beetle, *Circellium
bacchus*.' *Journal of Experimental Biology* 206 (2003): 1309–1318.

Byrne, Marcus, Marie Dacke, Peter Nordström, Clarke Scholtz and Eric Warrant. 'Visual Cues used by Ball-Rolling Dung Beetles for Orientation.' *Journal of Comparative Physiology* A 189, 4 (2003): 411–418.

Cambefort, Yves. 'Beetles as Religious Symbols.' *Cultural Entomology Digest*, 2 (1994): https://www.insects.orkin.com/ced/issue-2/beetles-as-religious-symbols/.

Cockerell, Theodore D. A. 'Dru Drury, an Eighteenth Century Entomologist.' *The Scientific Monthly* 14, 1 (1922): 67–82.

Coe, Malcolm. 'Dung: The Natural History of the Unmentionable.' *SWARA*, East African Wildlife Society (EAWLS), Vol 2, 3 (1979).

Cory, Alexander Turner. *The Hieroglyphics of Horapollo Nilous*. London: William Pickering, 1840.

Currie, Rolla P. and Andrew N. Caudell. 'An Index to Circulars 1–100 of the Bureau of Entomology.' *U.S. Department of Agriculture, Bureau of Entomology, Circular 100*. Washington: The Bureau, 1911: https://archive.org/details/circular1001100unit.

Dacke, Marie, Dan-Eric Nilsson, Clarke H. Scholtz, Marcus Byrne and Eric J. Warrant. 'Animal Behaviour: Insect Orientation to Polarized Moonlight.' *Nature* 424, 1 (2003): 33.

Dacke, Marie, Emily Baird, Marcus Byrne, Clarke H. Scholtz and Eric J. Warrant. 'Dung Beetles Use the Milky Way for Orientation.' *Current Biology* 23 (2013): 1–3: http://dx.doi.org/10.1016/j.cub.2012. 12.034.

Darwin, Charles R. *Journal of Researches into the Natural History and Geology of the Countries Visited during the Voyage of H.M.S. Beagle Round the World, under the Command of Capt. Fitz Roy, R.A*. Second edition. London: John Murray, 1845: http://darwin-online.org.uk/content/frameset?pageseq=1&itemID=F14&viewtype=text.

Darwin, Charles R. *The Descent of Man, and Selection in Relation to Sex*. London: Murray, 1890.

Dawson, Warren R. *Catalogue of the Manuscripts in the Library of the Linnaean Society – Part I. The Smith Papers: The Correspondence and Miscellaneous Papers of Sir James Edward Smith M.D., F.R.S., First President of the Society*. London: Linnaean Society, 1934.

Dongbin, Lü. *The Secret of the Golden Flower: A Chinese Book of Life*, trans. Richard Wilhelm. New York: Harcourt Brace & Company, 1970.

Emlen, Douglas J. 'Costs and the Diversification of Exaggerated Animal Structures.' *Science* 291 (2001): 1534–1536.

Emlen, Douglas J. 'The Evolution of Animal Weapons.' *Annual Review of Ecology, Evolution, and Systematics* 39 (2008): 387–413.

Fabre, Augustin. *The Life of Jean Henri Fabre*. Translated by Bernard Miall. New York: Dodd, Mead and Company, 1921.

Fabre, Jean-Henri. *The Sacred Beetle and Others*. London: Hodder & Stoughton, 1981.

Farber, Paul Lawrence. *Finding Order in Nature: The naturalist tradition from Linnaeus to E.O. Wilson*. Baltimore: John Hopkins University Press, 2000.

Findlen, Paula. *Athanasius Kircher: The Last Man Who Knew Everything*. New York: Taylor and Francis Books, 2004.

Forti, Luiz Carlos, Isabela Maria Piovesan Rinaldi, Roberto da Silva Camargo, and Ricardo Toshio Fujihara. 'Predatory Behavior of *Canthon Virens* (Coleoptera: Scarabaeidae): A Predator of Leafcutter Ants.' *Psyche: A Journal of Entomology* 2012 (2012): https://kopernio.com/viewer?doi=10.1155/2012/921465&route=7.

Fujitani, James. *Simple Hearts: Animals and the Religious Crisis of the Sixteenth Century*. PhD diss., University of California, 2007.

Garwood, Christine. *Flat Earth: The History of an Infamous Idea*. New York: Thomas Dunne Books, 2007.

Grayling, Anthony C. *The Age of Genius: The Seventeenth Century and the Birth of the Modern Mind*. London: Bloomsbury, 2016.

Griffiths, Tony. *Stockholm: A Cultural and Literary History*. Oxford: Oxford University Press, 2009.

Grubb, Wilfred Barbrooke. *A Church in the Wilds: The Remarkable Story of the Establishment of the South American Mission Amongst the Hitherto Savage and Intractable Natives of the Paraguayan Chaco*. London: Seeley, Service & Co., 1914.

Harari, Yuval N. *Homo Deus: A Brief History of Tomorrow*. London: Vintage, 2016.

Hawaiian Scarab ID. 'Scarab and Stag Beetles of Hawaii and the Pacific.' Accessed 8 May 2018. http://idtools.org/id/beetles/scarab/about.php.

Heinrich, Bernd and George A. Bartholomew. 'Roles of Endothermy and Size in Inter- and Intraspecific Competition for Elephant Dung in an African Dung Beetle, *Scarabaeus laevistriatus*.' *Physiological Zoology* 52, 4 (1979): 484–496.

Hornung, Erik. *The Secret Lore of Egypt: Its Impact on the West*, trans. David Lorton. New York: Cornell University Press, 2001.

Huxley, Julian. *Evolution: Modern Synthesis*. London: G.Allen & Unwin Ltd, 1942.

Immonen Esa-Ville, Marie Dacke, Stanley Heinze. and Basil el Jundi. 'Anatomical Organization of the Brain of a Diurnal and a Nocturnal Dung Beetle.' *Journal of Comparative Neurology* 525, 8 (2017): 1879–1908.

Jacobs, J. et al., 'First Come, First Serve: "Sit and Wait" Behaviour in Dung Beetles at the Source of Primate Dung.' *Neotropical Entomology* 37, 6 (2008): 641-645.

Jorink, Eric. '"Outside God, There is Nothing": Swammerdam, Spinoza and the Janus-Face of the Early Dutch Enlightenment.' *The Early Enlightenment in the Dutch Republic, 1650–1750: Selected Papers of a Conference held at the Herzog August Bibliothek, Wolfenbüttel 22–23 March 2001*, ed. Wiep van Bunge. Leiden: Brill, 2003: 81-207.

Jorink, Eric. 'Between Emblematics and the "Argument from Design": The Representation of Insects in the Dutch Republic.' *Early Modern Zoology: The Construction of Animals in Science, Literature and the Visual Arts* (2 volumes), ed by Karl A. E. Enenkel and Mark S. Smith, Leiden: Brill, 2007: 147-175.

Kalm, Pehr. *Travels into North America: Containing its Natural History, and a Circumstantial Account of its Plantations and Agriculture in General, with the Civil, Ecclesiastical and Commercial State of the Country, the*

Manners of the Inhabitants, and Several Curious and Important Remarks on Various Subjects Vol 1. London: T. Lowndes, 1770.

Keynes, Richard. *Charles Darwin's Zoology Notes & Specimen Lists from H.M.S. Beagle.* Cambridge: Cambridge University Press, 2000.

Livio, Mario. *Brilliant Blunders: From Darwin to Einstein – Colossal Mistakes by Great Scientists that Changed our Understanding of Life and the Universe.* New York: Simon and Shuster, 2013.

Losey, John. E and Mace Vaughan. 'The Economic Value of Ecological Services Provided by Insects.' *American Institute of Biological Sciences* 56, 4 (2006): 311–323.

Meyer, Isaac. *Scarabs: The History, Manufacture and Religious Symbolism of the Scarabaeus in Ancient Egypt, Phoenicia, Sardinia, Etruria. Also Remarks on the Learning, Philosophy, Arts, Ethics of the Ancient Egyptians, Phoenicians, Etc.* New York: Edwin W. Dayton, 1894.

Miller, Albert. 'Dung Beetles (Coleoptera, Scarabaedae) and Other Insects in Relation to Human Faeces in a Hookworm Area of Southern Georgia.' *The American Society of Tropical Medicine and Hygiene* 3, 2 (1954): 372–389.

Moczek, Armin P. 'Phenotypic Plasticity and Diversity in Insects.' *Philosophical Transactions of the Royal Society B: Biological Sciences* 365, 1540 (2010): 593–603.

Moczek, Armin P. 'Evolution and Development: *Onthophagu* Beetles and the Evolutionary Development Genetics of Innovation, Allometry, and Plasticity.' *Ecology and Evolution of Dung Beetles*, ed. Leigh Simmons and James Ridsdill-Smith. Oxford: Blackwell Publishing, 2011: 126-151.

Nicholas, Frank W. and Jan Nicholas. *Charles Darwin in Australia.* Cambridge: Cambridge University Press, 2008.

Petrie, W. M. Flinders. *Scarabs and Cylinders with Names.* London: Constable & Co Ltd., 1917.

Pizzo, Astrid, Angela Roggero, Claudia Palestrini, Armin P. Moczek and Antonio Rolando. 'Rapid Shape Divergences Between Natural and Introduced Populations of a Horned Beetle Partly Mirror

Divergences Between Species.' *Evolution and Development* 10, 2 (2008): 166–175.

Prest, John. *The Garden of Eden: The Botanic Garden and the Recreation of Paradise*. New Haven and London: Yale University Press, 1981.

Reynolds, Chevonne and Marcus J. Byrne. 'Alternate Reproductive Tactics in an African Dung Beetle, *Circellium bacchus* (Scarabaeidae).' *Journal of Insect Behaviour* 26 (2013): 440–452: 10.1007/s10905 -012-9365-1.

Ripley, George. *Compound of Alchemy. Liber Secretissmus*. California: English Grand Lodge, Rosicrucian Order, AMORC, 1993 and 2016.

Rösel, August Johann. *Insecten Belustigung*. Nürnberg: Johann Joseph Fleischmann, 1746.

Ross, Sydney. 'Scientist: The Story of a Word.' *Annals of Science* 18, 2 (1962): 65-85.

Roys, Ralph L. *The Book of Chilam Balam of Chumayel*. Washington, D.C.: Carnegie Institution, 1933: http://www.bibliotecapleyades.net/ chilam_balam/.

Sasson, Jack M. *Civilizations of the Ancient Near East*, Vol 3. New York: Charles Scribner and Sons, 1995.

Schreiner, Olive. *The Story of an African Farm*. Ed. Patricia O'Neill. Ontario: Broadview Press, 2003.

Shukla, Shantanu P., Jon G. Sanders, Marcus J. Byrne and Naomi E. Pierce. 'Gut Microbiota of Dung Beetles Correspond to Dietary Specializations of Adults and Larvae.' *Molecular Ecology* 25, 24 (2016): 6092–6106: http://dx.doi.org/10.1111/mec.13901.

Simmons, Leigh W., and Francisco García-González. 'Evolutionary Reduction in Testes Size and Competitive Fertilization Success in Response to the Experimental Removal of Sexual Selection in Dung Beetles.' *Evolution* 62, 10 (2008): 2580–91: https://kopernio.com/viewer?doi=10.1111/j.1558-5646.2008.00479.x&route=6Smith.

Smith, Kenneth G. V. 'Darwin's Insects: Charles Darwin's Entomological Notes.' *Bulletin of the British Museum* (Natural History) 14, 1 (1987): 1–143.

Smolka, Jochen, Emily Baird, Marcus J. Byrne, Basil el Jundi, Eric J. Warrant and Marie Dacke. 'Dung beetles use their dung ball as a mobile thermal refuge.' *Current Biology* 22 (2012): R863–R864.

Smolka, Jochen, Marcus J. Byrne, Clarke H. Scholtz and Marie Dacke. 'The Galloping Dung Beetle: A New Gait in Insects.' *Current Biology*, 23, 20 (2013): R913–R915.

Sutherland, John and Stephen Fender. *Love, Sex, Death and Words: Surprising Tales from a Year in Literature*. London: Icon Books Ltd, 2010.

Tribe, Geoffrey D. 'Pheromone Release by Dung Beetles (Coleoptera: Scarabaeidae).' *South African Journal of Science* 71 (1975): 277–278.

Tuxen, Sören Ludvig. 'The Entomologist, J. C. Fabricius.' *Annual Review of Entomology* 12, 1 (1967): 1–15: https://www.annualreviews.org/doi/pdf/10.1146/annurev.en.12.010167.000245.

Verdú, José R., José L. Casas, Jorge M. Lobo, and Catherine Numa. 'Dung Beetles Eat Acorns to Increase their Ovarian Development and Thermal Tolerance.' *PLoS ONE* 5, 4 (2010): 1 – 8: https://doi.org/10.1371/journal.pone.0010114.

Viljanen, Heidi, Helena Wirta, Olivier Montreuil, Pierre Rahagalala, Steig Johnson and Ilkka Hanski. 'Structure of Local Communities of Endemic Dung Beetles in Madagascar.' *The Journal of Tropical Ecology* 26 (2010): 481–496.

Walker, John Frederick. *Ivory's Ghosts: The White Gold of History and the Fate of Elephants*. New York: Grove Press, 2010.

Walsh, Benjamin, D. 'The Texas Cabbage Bug.' *Practical Entomologist* (1866): 1–10.

Ward-Harris, Joan. *More than Meets the Eye: The Life and Lore of Western Wildflowers*. Oxford: Oxford University Press, 1983.

Watts, D. C. *Elseviers Dictionary of Plant Lore*. Amsterdam: Elsevior, 2007.

White, Paul. 'The Purchase of Knowledge: James Edward Smith and the Linnaean Collections.' *Endeavour* 23, 3 (1999): 126–129.

Wood, John George. *Insects Abroad: Being a Popular Account of Foreign Insects Structure, Habits and Transformation*. London: Longman,

Green & Co., 1892 [1883]: https://www.biodiversitylibrary.org/
item/62694#page/9/mode/1up.

Yamada, Daigo, Osamu Imura, Kun Shi, and Takeshi Shibuya. 'Effect
of Tunneler Dung Beetles on Cattle Dung Decomposition, Soil
Nutrients and Herbage Growth.' *Japanese Society of Grassland
Science* 53, 2 (2007): 121–129: https://doi.org/10.1111/j.1744
–697X.2007.00082.x.

Index

Printed and bound by CPI Group (UK) Ltd, Croydon, CR0 4YY

27/10/2024

14580398-0004